By U-boat Through the World's Oceans

Paul Schultz

Volumes in the Great War at Sea Series

V188 My Torpedo Boat War Missions. Friedrich von Callisen
The Submarine War 1914-1918. Andreas Michelsen
The Last Cruise of the Majestic. George Goodrich
Two Lone Ships. Georg Kopp
U-boats Westward!. Ernst Hashagen
Six Years of Submarine Cruising. Johannes Spiess
German Submarines and their Warfare 1914-1918. Albert Gayer
U21 Saves the Dardanelles. Otto Hersing
By U-boat Through the World's Oceans. Paul Schultz

Cover artwork: Willy Stöwer's representation of the merchant submarine Deutschland's triumphant return to Bremerhaven from the USA.

By U-boat Through the World's Oceans

Paul Schultz

THE GREAT WAR AT SEA SERIES

Published by Trident Publishing
Miami, Fl USA

Library of Congress Cataloging-in-Publication Data

Originally published Im U-boote durch die Weltmeere
Leipzig 1931

By U-boat Through the World's Oceans
ISBN-13: 978-0-9963157-8-4
World War, 1914-1918 -- Personal narratives, German.
World War, 1914-1918 -- Naval operations.
Military operations, Naval--Submarine.

Printed in the United States of America

Translator's Introduction

Paul Schultz has left us an indelible image of, as he describes, "submarine driving" from the perspective of the affectionately classed Other Ranks, i.e. not a commissioned officer. Most submarine memoirs are by former commanding officers who see events unfolding via periscope or conning tower and execute the cerebral work of submarine duty. Stationed in the engine room, Machinist's Mate Schultz's work is physically arduous, occasionally leaving his mind free to consider the situation he is in and the unseen events unfolding outside the submarine. His world is one of reacting to commands.[1] Tending the engines and motors of a submarine was extremely tiring work due to the temperamental nature of the equipment, particularly the engines and in particular the engines on the boats he was stationed on and the concomitant working conditions.

In historical terms were are fortunate for Schultzl having served on three interesting submarines, U 21, U-Deutschland, and U 155. Each submarine and its commander is in some form a pioneer of German submarine accomplishments. U 21 gained fame for success off the Dardanelles in 1915, U-Deutschland as well for novel merchant submarine voyages in 1916 and U 155 operated as a U-Cruiser in 1917 and 1918, foreshadowing the long range submarine operations of the later period.

His memoir is chockfull of tales of adventure, humor, fear, tension, constant bone grinding fatigue and the unexpected. Paul's memoir is quite human in dealing with others; we see concern towards family, honor towards a fallen enemy, empathy for fellow humans, formerly enemies, in anguish when certain death in present. His

[1] The reader will note the differences in his experiences compared to a U21's celebrated commander's view as related in the submarine war diary of one particular tumultuous mission.

one liners punctuating experiences or introspective moments read modern 100+ years later: "You have to be lucky", "...[its] a dog's life", "It was just war, it was you or me.", "It was war and we were doing our duty" and "We imagined a different homecoming".

The work is at present the most interesting and genuine Other Ranks memoir. There is minimal patriotic talk and the text is immensely useful anyone interested in naval, submarine, Great War, etc. memoirs, operational, technology and social history.

As per custom, footnotes have been added clarifying or adding context to relevant sections. Events in the text have been verified with other sources where possible. Individual names from the vanquished are noted where possible keeping with the personal nature of the warfare experienced.

The text has been translated to provide the closest measure of the man, the author. Photographs from the original edition have not been included. There are no other omissions. Any errors in translation are entirely of my own.

Paul Schultz survived the war and faded from history. Fortunately he is very much alive in this work.

C.Corraliza

Forward

The submarine books published so far are almost exclusively written by officers. The expectation is justified that one day the report of a simple seaman will also find interest.

Any tendency towards showmanship has been dispensed with, because in the present booklet the only goal itself is the modest task of reporting how it actually was.

Paul Schultz

Oberliblar, April 1931

Table of Contents

By U-Boat Through the World's Oceans

I. On the submarine in peacetime

I fulfilled my duty as an active soldier in the years 1910 to 1913. The first garrison was Wilhelmshaven. There I was trained, initially in infantry service, which lasted about six weeks. After this training, my squad was asked who would like to go to submarines. We were given a cooling off period, but it didn't take long and we agreed to join the submarine service. Our squad consisted only of skilled craftsmen and when we entered the military we were already considered for submarine service.

Our new service began with theoretical training in submarine technology. First there was an electrical course, then a torpedo course. A final exam completed this first instruction. Afterwards we were sent to Kiel, our own submarine station. In another course we learned theoretical submarine construction. Using diagrams we gradually became familiar with the entire mechanism. For Example, although this new weapon was only just emerging, a submarine could best be compared with complicated precision clockwork. As there were only a few boats we had to wait a long time for an actual service command. Now and then we were taken to one of the boats present for an object lesson. During these visits we soon noticed our theoretical knowledge still needed to be complemented by a lot of practical experience and actual work is still the best teacher. The great hardships of the submarine service also became clear to us at this time. During the first visit my entire interest in this weapon began to wane. When entering the interior, I soon caught my breath because a sharp, pungent smell streamed towards me. The first boats were powered with kerosene.[2]

I had to come to terms with this inconvenience very soon when I received orders for U 14, kerosene powered boat. Since I was a

---

[2] Körting engines ignited as Diesel motors save for the fuel type.

trained mechanic and electrician, I was stationed in the engine room. I literally gasped for air, could hardly keep my eyes open and the acrid gases caused tears to roll down my cheeks. At the time, I didn't believe I would be able to endure for more than ten minutes in this compartment. When I peeked through a hatch to see whether the comrades in the next room were just as bad as I was, I was amazed to find they did their work unaffected and with the greatest possible ease. Of course, I didn't want to be left behind them and persevered. Still, the first night in this hell room was the most terrible I had ever experienced. After work was over I was so stunned by the terrible smell of petroleum that I immediately fell asleep like a half-dead man.

However, food on the submarine was very good. For this reason, many sailors preferred submarine service. The wages were also higher. In addition to wages, the crew received 1.50 marks for each diving day and another 6 marks a month for specialist allowances.[3] So submariners could afford a little more when they came ashore. The only unpleasant aspect for them was they smelled so much of petroleum because their clothes were soaked in the foul scent. I would have gladly waived the concessions, if only I had known how I could have been freed from this unpleasant service. Since I couldn't find a way out, I had to stay.

No long journeys were made by submarine in peacetime. During the short training runs, work consisted of torpedo shooting and diving attempts. Our training harbor was Eckernförde. The boats came in here in the evening and lay there during the night. After arriving, the boat was made ready for the next day, the electric batteries were recharged and work was carried out on the machinery, which often lasted late into the night. Then there was no end to the working day. Consequently the work was very exhausting.

During this time submarines were not allowed to do anything without an escort boat. A torpedo boat served as the escort boat.

---

[3] The average daily wage of adult male workers in Germany in March 1914 was 5.17 Marks.

One day on one of these training trips a man from the escort boat's engine room section fell ill. I signed up as a replacement and was ordered to serve on this boat. Initially I was stationed in the machine shop. The work wasn't difficult. But after a few days the NCOs asked me to wash their clothes and linen. I was to take on the role of a washerwoman on a part-time basis. With the best of intentions, I couldn't accede to this. Without further ado, I reported to the boiler room. However the chief machinist warned me, saying I shouldn't imagine this service too easy and it could only use very strong and untiring personnel. I explained to him I considered myself strong enough and promised to do my best. This is how I came into a torpedo boat's boiler room. The work was really very difficult and dirty. But I didn't care. I would rather have done this hard work than the unworthy one of a laundress. The first few days I had to trim coal, i.e. constantly place enough coal in front of the fire. The coal was stored alongside the ship's bulkhead in very narrow passageways. The coal bunkers on these little torpedo boats were so narrow you could hardly move in them. In addition, the intense heat of the boilers heated the coal so much it became dust dry. When shoveling you stood in a dense cloud of coal dust. You had to work in these rooms for six hours without interruption, and then you also had a six-hour watch. Of course, heavy seas made the work extremely difficult. If you didn't want to bump your head from the back and forth swaying, you had to hold on all the time, which was extremely tiring.[4]

After a brief job as a coal trimmer, I was ordered into the boiler room in front of the furnace. Although this work was one of the most difficult, I soon got used to it and the great heat. It was especially bad when the torpedo boat drove at high speed. Then you couldn't come to your senses at all. Whole mountains of coals vanished in a moment, more and more quantities were consumed

---

[4] Service on the mostly coal burning torpedoboats was so dirty and arduous the personnel so detailed were colloquially known as Schwarze Gesellen literally Black Tradesmen and considering sailor humour, with affection may be stretched to encompass Dirty Workers.

by the furnace, and shovel after shovel flew ceaselessly into the glowing throat.

With all the hard work, I was completely seaworthy in just a few weeks. I felt this especially during the autumn maneuvers, when we had to fight hard with the sea. Various stokers left due to the heavy swell. As replacements, we were joined by sailors who were only supposed to get the coal out of the bunkers. But after a very short time these personnel were exhausted. Rough seas and dusty air, that was quickly too much for them. We had no choice but to fetch the coals ourselves. These excessive efforts were not without consequences. When the watch was finished, it was difficult to drag yourself to the berth compartment. This was where the biggest mess was found. Everything was lying around in the corners and everyone was seasick. Due to the complete weariness resulting from the huge workload, your own resistance was of course greatly reduced, so that you were literally sickened in view of the bleak condition of the living quarters. You couldn't eat anything and so you just lay down with the others and tried to at least let your frozen limbs rest a little. Unfortunately, the hours of rest went by far too quickly, and far too soon it was time to relieve the watch. The digging began all over again, and we two stokers and two sailors did the same work that was otherwise done by eight to ten men.

Later I went from this torpedo boat to another, which replaced the first one. Even with the new boat, the sea voyages only extended to local waters, the Baltic and North Sea. After eight months of torpedo boat duty, I was able to return to the submarine service and enjoy a little more rest. I was back in the barracks for the last year of my service, even if my stay turned out to be less pleasant than I had hoped.

II. With U 21 to the Dardanelles

When war broke out I was 25 years old and in Koblenz when mobilization started. On the second day of mobilization, I had to report in Bonn. On the same evening we were transported to Wilhelmshaven in a freight train. The journey from Bonn to Wilhelmshaven took a full five days. The food we received these days was incredible. At every station where we stopped, the civilian population brought us so much to eat that it was impossible to eat it all. Well-filled ham sandwiches were thrown away or just the ham eaten away. With the best will in the world, it was no longer possible. Later this waste of food took bitter revenge. We only stayed the first night in Wilhelmshaven. The next morning we were brought to our parent unit and on the next transport went on to Kiel. There I first encountered a submarine barracks ship. After a few days this residential ship was brought to Wilhelmshaven, and we men formed replacements for transferred submarine personnel. Most of us were employed in the shipyard for work on submarines. Due to the small number of submarines, on-board service was not yet an option for us. But over the months more and more submarines were commissioned, and more and more improvements were made to their construction and, above all, to the machinery. We had to know all these innovations very well. The work process was by no means one-sided, as in a factory for example, but an extremely varied one for each individual and especially for those already trained. We had to familiarize ourselves with the electrical systems as well as with engines. Now the value of thorough peacetime training came into its own. Without this training it would have been impossible to have such a large number of well rounded men on hand.

But...as varied and stimulating as the work was, the desire actually arose in me more and more to go on a major war voyage in a submarine. Unsatisfied and longing for adventure, I often walked along the pier where the submarines were moored. Often enough I tried to get a orders to a boat, but always in vain. So it was in March 1915 when the word spread that a submarine was about to do something big. It was rumored the boat should go to Turkey. No one knew which boat was chosen for this.

On March 13, 1915, I was walking once again along the pier where several submarines were moored when a comrade from U 21 called to me. After the usual how are you? I asked my friend Arthur if they couldn't use anyone on board. I am tired of the work at the shipyard and would like to go on board on a trip. My friend now told me U 21 was to take a trip to Constantinople. I was very enthusiastic about this information. When friend Arthur continued to tell me they lacked a man for the diesel engine room, it was clear to me that under all circumstances I have to obtain this post on board U 21. We thought about the best way to make this happen. Since my friend also liked the idea of my coming aboard, especially since we had already served together in peace, we quickly agreed he would go straight to the chief engineer in charge. He went and Arthur reports he found the right man for the engine room. Of course, I went on deck right away. The chief engineer immediately came and asked me if I would like to go on the cruise to Turkey. Since I answered in the affirmative with the greatest zeal, he arranged for my immediate transfer. It took place the next day. Envied by all my comrades, I was finally able to get away from further peacetime work and go straight on a trip that promised adventure and danger.

On board I was stationed right in the diesel engine room. My first observation was the most gratifying because U 21 had no petroleum but crude oil engines.[5] This made the stay on board much more bearable. The petroleum engines had not proven themselves for wartime use as the exhaust gases betrayed the boat from a distance.

I settled in after a few days and was full of joy that my most ardent wish to take part in a cruise abroad was finally fulfilled. No work

---

[5] Diesel oil fired engines.

was too much for me. I was happy and willing to let the lofty mood affect me which had arisen due to the unique upcoming voyage. From the captain to the last man of the crew, the feeling prevailed we are about to accomplish something great.

We did some short practice voyages before we started the voyage. The 8.8 cm cannon we had on board was test fired. One day after these training trips had ended, the flotilla chief's torpedo boat appeared, docked with us, and the chief came on board. In a speech addressed to the entire crew of U 21, he drew our attention to what a great undertaking we were about to carry out, we had to be proud of having been chosen for this, and he expected us to all do our duty. After the hurrah, he said goodbye to everyone with a handshake. We were really proud then. We ought not only to accomplish a great military deed, but above all to give testimony to the whole world that German technology was actually at the forefront.

Departure was delayed when our compressor failed and the boat had to visit the shipyard again. But otherwise the final preparations were finished; supplies and provisions were taken on board. We were also allowed to write to our relatives but we were strictly forbidden to mention the destination of our trip.

If in peacetime a submarine was not allowed to go on an unaccompanied cruise, it had all changed during war. Submarines had to make their journeys alone and take on all the dangers. Previous successes had shown they were able to achieve great deeds of the first order. But all these successes had been achieved in a relatively narrow area. The latest goal set for the submarine surpassed everything achieved so far. Such a long journey had been generally regarded as an impossible undertaking, and there were not a few who were convinced of our certain demise. We too knew only too well no submarine had undertaken such a long journey with so much danger, but the whole crew was full of confident hope.

On Sunday, April 25th, 1915, we started the great journey in our tiny boat. In the morning we awakened at 4.20 a.m., U 21 cast off at 5.30

a.m. and went to sea. Our departure took place quietly and unnoticed, so nothing of our plans would leak to the public and our enemies would not find out about it. Submerged from 9.15 a.m. to 9.50 a.m., the boat made a trim attempt, i.e. if it was balanced underwater. We were able to cruise calmly and undisturbed all Sunday. No mast could be seen of sea-ruling England. On the first day of our voyage, the sea seemed to be "empty" of traffic. Only on Monday did enemy vessels come into view. Not just one steamer, but two destroyers. Before they could see us, we submerged at 12:45 p.m. and stayed invisible for an hour. But we did not get to rest on Sunday. At 5 o'clock the alarm bells rang again when several enemy steamers appeared. But we already had the feeling of absolute security. All the procedures necessary for diving were done calmly, albeit with great speed. None of our enemies could see us. Contacts were repeated on Tuesday. But we always dived in good time so no opponent could spot us. On Wednesday we turned around the northern tip of England and now - driving around Ireland in a wide arc - we headed south. We went into hiding if we saw an enemy vessel. Our trip was supposed to remain secret and unnoticed.

Unfortunately, we also suffered a mishap that almost turned out to be fatal. We had very rough seas, such that the sea came onboard and found its way into the engines' exhaust. Seawater penetrated the engines themselves, where it evaporated and left a hard crust of salt. This crust became so thick the engines failed and stopped. Both engines now had to be dismantled and completely overhauled. All available staff had to help. Work continued day and night without interruption to repair the damage. To make matters worse, the sea was constantly so stormy you could hardly stand on your feet. The work lasted from Tuesday to Friday afternoon. When it was over, everyone involved was utterly exhausted.

During these days we were dependent on the power of an electric battery, which could only be conserved by allowing us to drift with the current. Fortunately, we didn't have to dive. Had the conditions forced us to do this and possibly forced us to stay under water for a lengthy period of time, the battery would have been exhausted and we would have been hopelessly lost, as we had no way of recharging it. But you have to be lucky, and this disturbance passed without any serious consequences. But on the continuing voyage, all the more care was taken and everyone tried to protect the boat from the slightest damage.

From now on it was a brisk ride south. At a good distance from the French coast, we saw only a few vessels, from which we always hid ourselves in good time. We were spared from storms, only now and then we had a little choppy sea. Nevertheless, the technical section's work was extremely exhausting. There was little rest. However, six hours of duty and six hours of rest were planned. Very often, however, the alarm went off when one had just finished six hours of duty. The boat had to e.g. suddenly dive. It was a matter of course that one then had to stay at one's post, at least as long as the dive lasted. But since we often stayed underwater for six hours or more, the watch was just as often extended to twelve and even eighteen hours. Repairs often had to be made in addition to the usual work. So there was no real sleep. Undressing was also prohibited during rest hours. We were forced to always wear our oil-soaked clothes. There was also no water for washing. Existing stocks had to be used sparingly, as one did not know how long we were to be underway and what was still in store for us.

Due to the smooth and undisturbed journey along the French coast, we were already near Spain on May 1st. On Sunday May 2nd we stayed near Cape Vilan[6] during the day. We watched an English cruiser pass a long way off. We noticed our boat was slowing down, but we couldn't find any explanation for this. Our astonishment was therefore enormous when we learned that U 21 was expected at Cape Finisterre by a Spanish steamer in order to be supplied with fuel and provisions. We should meet the steamer in the dark. From dawn until late at night we cruised back and forth, always on the lookout. I had just been on watch and was on deck when I suddenly saw a dark mass heading towards us. A signal lantern flashed at short intervals and then went out again. We received the signal to

[6] A peninsula on the west coast of Galicia, Spain. A navigation landmark for shipping.

follow after signaling in reply that we recognized the steamer. We docked next to the steamer in a quiet, dark bay and immediately began to work. We noticed the crew of the steamer were Spaniards, except for a German gentleman, from whom we probably rightly assumed that he was acting on behalf of the German naval authorities. Obtaining the steamer and the crew for our purposes may well have cost a fair amount of money. Taking over the cargo turned out to be difficult. It rained continuously and the sea was very choppy. Completely soaked, we worked until 4 o'clock in the morning, with the choppy sea constantly pouring its splashes over us. From some distance in the bay the light rays of a brightly lit German steamer came over to us. We learned that he had fled here at the outbreak of the war and found shelter. Some of the provisions might even come from him.

Since we had not finished the transfer in the nighttime from Sunday to Monday and could not carry it out during the day, a new meeting was arranged with the steamer for 9:00 p.m. But during the day it was found, to our greatest regret, that we couldn't use the oil that had already been taken over because it was heavy tar oil[7] and could not be used for our engines. So all the work involved in taking over the oil had been in vain. We had no choice but to pump the fuel which had been brought in with so much effort into the sea. When we met the steamer again that evening, we could only tell him about the misfortune. Our friends were very sorry about this. Of course, any further intended takeover of oil was now unnecessary. We therefore said goodbye to the steamer and its crew with warm thanks for the help and support they brought us. Everyone wished us a good journey and good luck.

When we discovered the oil taken over was worthless for us, the boat's leadership was faced with a very difficult decision. Captain Hersing[8] and his chief engineer considered what to do, whether to continue or return. On the basis of maps it was determined the way

[7] Crude oil in modern parlance.

to a German North Sea port was as long as to the port of Cattaro (Austria). With the most economical consumption of supplies, especially of fuel, we would be able to reach one of the ports. Captain Hersing could now easily determine which way we were to go. It was a proof of his excellent leadership that he had the whole crew come together, explained to them how things stood, and then raised question as to who would be willing to continue the voyage. This discussion showed the best expression of comradely cooperation. Of course everyone agreed to continue. If we had already come this far, then we wanted to try to achieve the prescribed goal. Hersing also said in an emergency we could call at Italy. Of course, the journey now continued.

We passed the coast of Portugal, but in such a way for us not to be seen, we headed for Gibraltar without any special incidents. We had not yet reached it when we heard via radio that Italy had declared war on Austria. Any possible stop at Italy was now removed from our calculations. Circumstances required practicing the greatest economy on the remainder of the journey; especially if we wanted to reach Cattaro. Lighting use was conserved. Submerging should only be done in extreme emergencies, because the battery was needed for this. It was to avoid recharging the batteries too often, which supplied power to the electric motors, because this entailed considerable oil consumption. Every day oil consumption had to be reported and every day new calculations were made as to whether supplies could still be sufficient.

We noticed from the constantly increasing temperature that we were moving ever further south. On Wednesday evening May 5, 1915, it was announced we were only 120 nautical miles from Gibraltar. If conversation on board had previously been geared exclusively to the coming events, it now became even more interesting given the proximity of Gibraltar. The question of passage through the strait was discussed in detail by all ranks. Some were of

---

[8] Kapitänleutnant Otto Hersing,(1885-1960), well regarded submarine captain who in September 1914 had achieved the first sinking of a vessel by a submerged submarine launched torpedo.

the opinion the passage would bring us much more unpleasantness; others feared we would be discovered and followed, forcing us to often go underwater. This would put too much stress on the fuel supplies and make it difficult to continue driving. A third group, and indeed the most numerous, was firmly convinced that considering the course of our journey thus far, we should continue to count on good results. Through his skillful leadership, Captain Hersing had shown if it were necessary, he would steer U 21 through the clutches of the devil. We all had unlimited confidence in our leader.

Politics was not discussed at all. All conversations were about our trip, as we were all only imbued with the one wish to have as much opportunity as possible, to bring real help through deeds to the beleaguered fatherland and to struggle as much as possible to free them from the clutches of our many enemies. The festive atmosphere which had prevailed since the start of the trip continued undiminished, and everyone was proud to be part of U 21's crew.

As a precaution, a doctor been added to the ship. But he was only seen by everyone as a so-called bather[9], because no one was allowed to get sick. The naval doctor has had little opportunity to show his art, apart from a few small things, such as for injuries and ailments caused by climate change, which were cured with the panacea aspirin.

On Thursday, May 6th, at 2.30 a.m., we sighted a beacon on African north coast. 6.30 am we drove into the Strait of Gibraltar. We were able to safely proceed surfaced until noon. But suddenly at 2.30 p.m. the order to dive quickly rang out. Three enemy destroyers chased our boat. No sooner had we dived than we were shot at, although a few seconds too late and without our being hit. The task now was to evade the chase by skillful underwater maneuvering. Because it had to be reckoned as fact that we - now sighted and recognized - would be pursued with the greatest zeal and fierceness

---

[9] Friendly, derogatory term for a guest aboard without seagoing experience.

in order to render ourselves harmless and to destroy the ship. But our Hersing masterfully succeeded in deceiving the pursuers and evading them, so the next day we reached the Mediterranean Sea without being disturbed.[10] We did not remain undisturbed for long. On Saturday we were sighted by an English steamer and pursued until 5 p.m.. Hersing also evaded this opponent with great skill.

After crossing the Strait of Gibraltar, pursuits in the Mediterranean became extremely dangerous. We repeatedly sighted war and merchant craft whose purpose and task was to hunt down and destroy U 21. Time and again, Captain Hersing managed to escape the pursuers. On Sunday, May 9th, in the afternoon at 4.15 p.m., quick dive[11] when two torpedo boat destroyers were seen in our immediate vicinity. 6.20 o'clock quick dive again because of a steamer that was apparently on duty as a guard vessel. In the evening at 10.30 a large cruiser passes us some distance away. Since we have completely dimmed the boat, he cannot see us. Our ship moves quietly and mysteriously through the waves, striving towards its goal like a ghost ship. Only at the front of the bow, where the waves break, is there a silver streak.

As a result of the favorable night journey we were on Monday May 10th at the southern tip of Italy. At 4 a.m. early in the morning we sighted five destroyers at a distance of about 500 meters. They too seemed to have the task of hunting us down. [12] They searched the sea with their searchlights, but without getting their cones of light onto us. For us, especially for our captain, the presence and activity of five destroyers meant we had to increase our attention and our willingness to act as quickly as possible. Any carelessness could mean doom; any moment could bring us a devastating projectile,

---

[10] Lt.-Cmdr Wilfred Ward Hunt (1883-1970) commanding Torpedo Boat T.B. 92 of the Gibraltar Patrol write he nearly rammed U 21 and was so close it would have been destroyed if depth charges had been available at that time.

[11] Schnelltauchen in German. Being an oral command, vocal inflection is an indicator of urgency and/or need for more urgency.-quick to fast/faster. See Schultz's explanation in the following page.

[12] Entente naval forces had been tasked with patrol possible navigation landmarks and crossing points to catch U 21.

any moment we could be rammed by one of the vessels. In the face of the terrible danger, every man had to be utterly focused on duty. Only in this manner were we able to successfully reach the evening darkness from Monday to Tuesday, which gave us a little more protection again.

Throughout Tuesday we drove on using the greatest caution without any untoward events. Suddenly at 2 a.m. we were surprised by a destroyer. A steamer in his company immediately tried to ram us. We received heavy fire from the destroyer. This time we really seemed lost. But Hersing's orders to dive quickly were carried out with such marvelous swiftness that we were gone like lightning from the scene. Just a few seconds' delay and we would be done for. Underwater we heard the steamer rushing overhead at top speed. We had never dived as fast as on this occasion. Of course we reached the safe depth for us of 25 meters in no time. Only here did we feel safe. The bombardment cessation led us to assume that, as a result of our lightning-fast disappearance, the two vessels expected to have destroyed us. In our safe depth we heard the pounding engine noise of the ships searching over us for some time. We had escaped the greatest danger which had threatened us so far.

This quick dive which saved us so often was always a daring step given the state of submarine technology at the time. You have to imagine it. Suddenly, while doing the usual work or even while sleeping, the bell rings through the whole boat: Quick dive! Everyone who hears this bell signal passes it on by shouting loudly. The whole crew is shouting "quick dive" at the moment. Like a scream it goes through the whole boat. Those who are asleep jump up and rush to their posts, and the call continues: quick dive! This scream, mixed with the roar of the machines, increases to a real hell of a noise, terrifying to the innermost nerve, it goes back and forth on board until everyone is at his post. With the inevitable rush, simply worried about getting to his post as quickly as possible, someone hits a bump on his head here, there someone gets a blue shin because valves, hand wheels and protruding valves are in the way everywhere. Crawling through the hatches connecting the bulkheads certainly does not promote the necessary speed. They are only so large that a stooped man can barely pass through. All of this makes quick diving a stressful business time and time again. If there is even one of the crew who loses calm in these moments and is not fully alert and attentive to his post, then the dive can very easily go amiss. Just one wrong push of the lever, a single wrong turn on the countless wheels and switches, would cause the boat to sink. Everything as in the case just described has to often be done frequently with lightning speed. So diving itself is always associated with great danger. In addition, even with infallible crew reliability, equipment can fail, and what was more, in our special situation, it was extremely unpleasant that we had to be careful with the fuel supplies.

The next day we were able to continue our journey without any particular incidents and came quickly closer to our goal. Once the Strait of Otranto had been passed, further travel through the wonderful Adriatic Sea was more of a pleasure trip. We were able to continue our journey safely surfaced on May 13th. We finally spotted a destroyer. Due to the experience we had on the night of the 11th to the 12th, we were extremely cautious and at first observed the destroyer with the greatest suspicion. On getting closer, however, it was found he was flying the Austrian flag. Captain Hersing signaled identification. A signal was sent back from aboard the destroyer indicating he had orders to lead us through the mine barrier. We followed him and made it safely to the port of Cattaro. A wonderful scenic picture lay before us. The tall Montenegrin mountains all around. The Lovcen[13] towered above all others and was clearly visible in its majestic size.

In Cattaro we were warmly welcomed by our Austrian Brothers. Our arrival had been announced, and everything that could be done was ready for our reception and welcome. We were cheered with incomparable joy. Everyone wanted to see the heroes of this outrageous journey in a tiny submarine and express their fine admiration.

---

[13] Mount Lovćen, 5.738ft. overlooks Cattaro.

We were assigned the barracks ship "Gäa" as quarters, which lay in the middle of the harbor. Exhausted as we were all from the exertion, we looked forward to being able to sleep in. Our U 21 was to be completely overhauled. The diesel engines were subjected to a particularly thorough inspection and test. All the work was done by ourselves, even if we had to work significantly longer than usual. But we wanted to do it. We knew by now from experience and the dangers we had overcome that it was absolutely necessary to always have our boat in the best possible condition if we wanted to continue to achieve our ultimate goal. So far we have avoided every engagement in order to reach our actual battlefield, the Dardanelles, and because of the coming events we thought of bringing our boat to peak performance. That is why all of the work, even the slightest ancillary work, was carried out with the greatest care. The only unpleasant thing about this overhaul was the great heat prevailing in the port of Cattaro and to which we were not yet used. Otherwise the stay was a real recovery. After few days we got leave in the afternoon. Austrian comrades, proud of us, gladly accompanied us on our visits ashore and led us among the very beautiful surroundings of Cattaro. During these excursions we learned no one could believe we had made the trip to Cattaro entirely by water and through the Strait of Gibraltar. It was generally assumed our boat had come by train to Pola and was assembled there.[14] Repeatedly our allied brothers urged us to tell them the truth. It was really not easy to convince them of the fact we traveled by sea. But when they finally realized we had actually achieved what seemed to them to be incredible, their fraternal feelings reached the level of highest admiration.

On our walks we also met many German troops who were on the march to Lovcen. Transport to this Montenegrin mountain fortress was extremely difficult, as only donkeys and a particularly suitable breed of horse could be used. The Austrians tried several times to storm the fortress, but they hadn't succeeded. That is why German

[14] Smaller submarines were indeed shipped disassembled from Germany to Pola for reassembly.

troops were called in. Germans did not have to step in everywhere when the Austrians were making forward progress.

Our stay in Cattaro lasted exactly one week. The repair work on U 21 was done to the extent that we could continue our journey to intervene in the events at the Dardanelles. Once again, trim and dive attempts were made in the harbor. In the afternoon of May 20th we got a few more hours of leave. Everyone had to be on board at 6 o'clock. Since the time of our departure was announced, virtually everyone appeared again for our farewell. When we cast off at 6.3 o'clock, incessant hurray and cheers of well being roared after us. The waving lasted until we were out of sight. With the best weather we reached the Adriatic again. At 9 o'clock in the evening we had to dive quickly again for the first time because of an enemy vessel.

The next day we passed the island of Corfu at a greater distance. The trip was pretty undisturbed. We only sighted an enemy destroyer where we dived early enough not to be noticed. As we were getting closer and closer to the actual goal, we had to proceed very carefully so the enemy did not find out we were on our way. We tried to conceal ourselves from sight as soon as we saw any vessel. Over the next few days, the number of enemy seekers increased. In order not to be in danger inadvertently, we drove submerged during the nights.

On Sunday, May 23, 1915, the first day of Pentecost, we saw no vessels and were therefore able to drive towards our goal at the fastest possible speed. We left several clearly visible Greek islands at a considerable distance from us.[15]

[15] U 21's war diary describes the journey from Cattaro to the chapter end date:

May 17- 18 Cattaro Mouth	Oil transfer. Transfer must take place from barrels. Procedure as inCorcubion. Duration of the transfer two days.
May 20, 6.50pm	Received orders to sail. Castoff from S.M.S. "Gäa".
May 20 NWAdriatic Sea	Piloted by S. M. Destroyer "Pandur" through the mine area.
May 23	The route mentioned was chosen so as not to cross any shipping lanes

---

Aegean Sea during the day and not to pass any of the large inhabited islands too close. Tomorrow I intend to go around Lemnos to the north, going west, first to observe the position of the enemy ships and possibly use the twilight or moonlit night to attack.

May 23 1145am A warship ahead in sight with two masts and several funnels signaling a steamer standing under land and is apparently investigating it. Submerged. The ship is recognized as the Russian cruiser "Askold" (5 funnels) as it approaches. The ship is at anchor south of Dedeagatsch approx. 5 nm from land.

2pm Course 120 ° I decide not to attack the Askold, although the shot seems safe. Instead, I intend to proceed submerged to Gallipoli first in order to reconnoiter the position of the Anglo-French fleet and to commence wholesale the first attack there. Since I can expect to hit the fleet without any news of my approach, I will announce that by attacking the non-English cruiser, which is only 6000 tons, and the prospects of later attacks will be reduced.

4pm Arun class destroyer B.B. ahead in sight with a westerly course (apparently) towards Askold.

450pm Course 140 °, steamer with sailing vessel in tow at B. B. in sight with westerly course. Cannot yet surface.

555pm Surfaced. Stayed and charged in place. Various vessels in sight heading west, which are avoided.

950pm A dimmed vessel, apparently a destroyer, approaches the boat.

III. Against the enemy

On the second day of Pentecost we drove towards the Dardanelles. It is in this area where we should come to action. The mood on board was completely different. The previously festive joy had given way to a cold and sober determination. Everyone was ready to give their last in the fight with the enemy. In the afternoon there was another quick dive because of a smaller vessel that shouldn't be attacked. From Monday evening at 10 p.m. to Tuesday evening at 9 p.m. we proceeded under water. During this submerged voyage we sighted the English battleship "Triumph" on Tuesday morning. We approached cautiously, waiting for the most suitable moment to attack. The crew got the torpedo ready to fire and waited for the command. Everyone was filled with feverish tension. Now was the moment for the sake of which this journey of life and death had been undertaken, now was the moment in which all the hopes and wishes with which we had been sent from our fatherland had to be fulfilled, now it had to be shown whether we too would add a new exploit to the previous ones by submarines. With every nerve strained, everyone was at his post to carry out every command with the utmost speed and accuracy. Minute after minute passed. Only cautious maneuvers were carried out in order to get close enough to the enemy and yet go unnoticed. With the smooth and calm sea it was an unparalleled risk to sneak up to the giant ship in the small boat, for whose protection a number of destroyers were also ready. None of us had any doubt we were hopelessly lost if we were noticed prematurely or if the first shot failed. But none of us harbored the desire to wait for a more favorable sea for the attack.

Finally - after hours of back and forth it was time. Captain Hersing gave the first order to fire. The attack took place at exactly 11.30 a.m. Everyone in the ship held their breath. The first shot! Will it hit? Will it hit so accurately it immediately destroys the enemy completely and there is no time for him to look for us and kill us instantly with his guns? All the questions were quickly resolved. A loud crash gave us the answer. A hit. In addition, as it immediately turned out, with telling accuracy. After the torpedo exploded in "Triumph", it lay on its side and immediately sank. He was U 21's first victim and with him 12,000 tons and the entire crew sank into the depths.[16]

The tremendous detonation of the torpedo had been noticed by neighboring ships, and they rushed to the scene of the disaster to help. We would have loved to have made further attacks. To our greatest regret, we had to do without this because the sea was too calm and any movement would have been immediately noticed. Despite our great caution, we must have been spotted because we were suddenly shot at and various vessels tried to ram us. The only way we could evade this attack was to seek depths beyond our pursuers' reach. It was only at 9 o'clock p.m. when we dared to surface again. The long underwater voyage had almost completely drained the electric battery. It had to be replenished by the diesel engines, and only now did we get excited about the first great success. The Gulf of Saros, in which we sank the "Triumph", has remained unforgettable to all of us.

The next day we looked for new opponents. But there were no longer any larger warships to be seen in the bay where the English

---

[16] U 21 had sunk other ships prior to this vessel. Reference is to sinkings achieved on this special mission.

An officer onboard HMS Triumph reported: "... we saw the periscope on the starboard side and the trail of torpedo right after at 12.30. Instead of bursting in the nets as we expected, it went through as if there was no net there and hit us in amidships. Triumph lay 20 degrees on the starboard side and it was impossible for us to take our lifeboats. Our list quickly deteriorated, and after a few minutes it became clear that nothing could save the ship, so the order came to abandon ship. A destroyer came to the port side, but the net defense made things difficult because it was impossible to create openings on the port side, except for the upper deck and stern deck. For myself, I jumped from the upper deck and swam away from the ship. Ten minutes after the torpedo hit, the ship capsized and remained upside down for a while, eventually gradually disappearing... " Also, the majority of Triumph's crew was saved. Most sources indicate 73 lost, over 500 survivors. U 21 dived immediately following and was not privy to the rescue effort.

fleet had been based. They preferred to go to another hiding place. There too they were found by us.

On Thursday, May 27th, an English battleship, "Majestic", was sighted. It was in one of the bays of Sedil-Bahr. This time it was even more difficult to get to the opponent. The battleship lay behind a line of smaller steamers that surrounded it in a semicircle to protect it against a submarine attack. Recalling the sinking of the "Triumph", the enemy wanted to secure their capital ships from attack if possible. It was therefore much more difficult this time to get to the opponent. More so since the enemy was now alerted and kept a careful watch. The enemy had not been too vigilant before the attack on the "Triumph". One had expected the devil in person rather than a German submarine. But now they were on the lookout knowing this plague was also making itself felt in the Mediterranean.

The danger was therefore much greater; the chances of success were lower. Yet there was no hesitation and no wavering. Just as every single man had only one thought of destroying this enemy, so did our Captain. He dared to attack with as much prudence as boldness. He approached with the utmost caution. It took an hour and a half to get close to the giant, and for an hour and a half the crew of U 21 was strained to the last nerve. At 5.30 a.m., Captain Hersing was finally able to give the order to fire. Another minute of extreme tension. Our position was more vulnerable this time unless the shot had an immediate devastating effect. Once again the excellent training of the entire crew proved its worth. This time, too, the fired torpedo was a direct hit. The battleship "Majestic" sank so quickly that we were able to surface again after 40 minutes.[17] When

---

17 On board HMS Majestic Petty Officer J,G. Cowie recounts the event after alerting an officer to the presence of a periscope, who replied "'Yes, and here comes the torpedo.' The commander shouted a warning to all the men on the port side, and I shouted down the ventilator leading to the engine-room. The torpedo was traveling at a good rate, and I saw that it was going to strike somewhere forward, so I went to the end of the bridge to see the effect.

It struck! There was a great muffled roar and the old ship quivered and shook in a terrible way. The masts and yards swayed as if they were coming down on top of us, but they held good. A huge volume of water shot up for about 200 feet in the air, on

searching the place of the sinking we saw no survivors, so we assumed that the entire crew of at least 1000 men had also sunk.[18]

The loss of two large battleships within two days had an unexpected effect. All vessels left the Dardanelles entrance. The ships were kept in safe harbours when it was finally understood they were dealing with a German submarine. The islands of Lemnos and Imbros became the bases of the fleet. Further activities were directed from here. Only destroyers were sent out to find and destroy us, and a large reward was offered for our destruction.[19] From now on, we did not see any larger warships.[20]

Due to the terrible tension, we had a well-deserved rest day at sea on Friday. Only charging the electric battery was done. Suddenly at 11.30 a.m. we were surprised by a destroyer. Again an emergency quick dive. We proceeded on again in the evening. Throughout the night we drove again underwater and kept ourselves carefully covered from sight the following morning. We constantly searched for enemy ships. We vainly searched some bays along the heavily indented coastline. There - suddenly luck seemed to be on our side again. A modern warship was spotted in a hidden bay. Since the sea was slightly choppy, Captain Hersing considered the opportunity for an attack to be favorable. It was prepared with the greatest care and the boat was brought up to a distance of 800 meters from the ship. I stood in the central hatch to watch the torpedo running time

---

the port side, and I knew that our end had come. Things began to move very quickly then. Men were running in every direction, but very quietly and swiftly, for they realised that we must fight for our live, and that at any moment an internal explosion might ruin our last chance. The order was passed 'every man for himself' and everything that would float was being commandeered as a possible aid to rescue. Everyone seemed very cool, and there was scarcely and shouting... Tugs and trawlers were already coming to our assistance, but they couldn't come alongside as the ship had heeled over a lot to port, and it would have been dangerous for them. There was only one thing to do, and that was to swim for it."

[18] The majority of Majestic's crew was rescued while U 21 submerged to evade counterattack. Most sources indicate 40 casualties; 737 survivors.

[19] A rumored GBP 50,000 was offered.

[20] British battleships were withdrawn from bombardment duty after the double sinking. The work being carried out by lighter vessels with smaller caliber artillery.

on a clock. The command: Shooting![21] Hersing had already given. We just waited for the execution command: Go![22] But this didn't come. Instead, the captain ordered: Everyone to the stern! We're in a submarine trap, in a net! The next order came immediately: both motors emergency power astern! This order was carried out, but it was unsuccessful. We were stuck. A tremendous agitation took hold of the crew. We didn't expect a submarine trap here. It was soon discovered the net had spread over the whole boat and was stuck to the many protruding parts of the boat. Anyone who felt like it could see the mess through the periscope. Of course, I did not miss this opportunity. What could be seen was not very pleasant. The net, made of rather strong steel cables, had wrapped itself around our boat like a cobweb. After this bitter statement, the fateful question arose: What now? There was little hope of getting out of the trap. At any moment we expected some enemy vessel would appear to finish us off. We wanted to avoid this fate at all costs. With determined energy we set out to tear the net apart. We hoped to free ourselves by forwards and backwards use of the motors. We worked in feverish excitement for hours. We foresee what came next. The boat's ability to move freely was almost exhausted due to the long underwater journey and heavy use of the accumulator battery. The chief engineer dutifully reported to the Captain that the batteries would soon have nothing more to offer if the motors were to continue to be used in the previous manner. The air became increasingly unbearable; as did the prospects of liberation by one's own efforts. Our fate seemed sealed. While pondering this situation the call suddenly rang out from the command tower: "I will give the final command!" Anyone can imagine how we felt here. The command was: port engine emergency power astern and starboard engine emergency power ahead! We were in an indescribable tension as to the outcome of this order's execution. Life or death depended on success or failure. Into this tension penetrated the

[21] Zum Schuß! In German. Literal "For/To the Shot" implying last look prior to the torpedo launch command.

[22] Los! in German. Literal or colloquial derivatives indicate a release action: "Go", "Away", "Release", etc. Not "Fire" as this was reserved for an actual onboard combustion and treated as an emergency instead of a command.

sound of the net tugging and tearing. The boat trembled in every joint under the various effects of the engines. Suddenly a jolt and the voice of our incomparable leader rang out: We are free! The net is torn. These words must of course have a profound effect. They created in the soul of each of the crew as great a sense of release as the tension had previously. We turned slowly and headed for the bay's exit. We were only able to free ourselves thanks to the strong seaway. If the sea had been less restless, the English would have noticed us from aboard their ship by the net moving back and forth. In the best case scenario, they would have pulled us up with the net and towed us away. Probably, however, for their own safety, since they had to reckon with an attack from us, we would have been subjected with a devastating explosive shellfire. So we had escaped a danger, where with a 99 percent probability, should have been our end. Who knows how we would have fared had we not had our commander Hersing, with his indomitable will to do extraordinary deeds and his rare ability? We owed our rescue primarily to his daring and his precise knowledge of the U 21's capabilities.

After we were able to move freely again, the order was given for all available personnel to arm themselves with tools in order to free the boat from the remnants of the net immediately after the next instance we surface. Parts of the net could easily impede the boat's mobility and maneuverability. The order was scarcely given when everyone tried in great haste to get possession of a suitable tool. As soon as we surfaced, the designated personnel hurried on deck to get the job done as quickly as possible. Every moment we had to reckon with encountering a new danger. We immediately started to dismember the net components which were still hanging on the boat. That wasn't done as quickly as I mentioned. The wires were very strong steel cables were soldered with lead at their junctions. While the liberation work was being carried out on deck, the boat's batteries were quickly recharged so we were again fully equipped. At last we were ready to continue our journey. That night we stayed under water again. The next day we returned to the place that had almost become disastrous for us to see whether the warship was still there. It should be rewarded for the inconvenience it caused us. Unfortunately it was gone.

Due to the various events, not all of which were foreseen in our planning, we had to linger longer than intended and food became scarce. As a result, we felt compelled to seek land as soon as possible. We didn't have to wait long. On Tuesday, June 1st, we came to the somewhat hidden bay of Budrum, where we were expected by a German Lieutenant from the Göben and a few Turks. Here we docked, greeted with admiration by the small command. Our deeds were now known. The lieutenant had orders to lead us through the Dardanelles' minefields. In the meantime our provisions had run out completely. We had neither bread nor potatoes on board, let alone meat. To greet us, the Turks slaughtered some goats and roasted them on a spit. Since we longed for a piece of fresh meat, we looked forward to the impending consumption. When the Turks thought the meat was ready, the goat was cut in pieces and each received his piece. But oh the fright! The meat was so tough and still quite raw we couldn't enjoy anything, absolutely nothing of it. We definitely had to deny our new friends any culinary skills. They had meant well, but probably thought we were savages, for which our appearance gave them every reason. The whole crew had washed twice in almost three weeks. Of course we were all unshaven too, and you can imagine how we looked. When the lieutenant found out we had no more food on board, he immediately instructed several of his men to fetch something for us in the nearest towns. After a few hours those sent out came back heavily packed. Bread, eggs, meat, a lot of fruit, especially oranges, which we were particularly welcome in southern temperatures; we received such abundant quantities we could very easily reach Constantinople with the supply.

The next morning at 9 o'clock, after we had slept undisturbed for a few more hours, we cast off and set course for Constantinople. We proceeded again with the utmost caution, as the area was heavily patrolled by enemy vessels. In the afternoon we had to dive because of an enemy steamer in order not to be spotted. On Thursday we dived again and stayed underwater all night.

The English spied us although we were taking every possible precaution. On Friday several enemy destroyers followed and shot us. We were also chased by an enemy aircraft. Since the destroyers attracted fire from the land batteries from the German and Turkish positions, we were in double danger. Splinters from projectiles often fell on our boat. We assumed an airplane was pitching explosive charges at us. As the bombardment became increasingly heavier, Captain Hersing gave the order to go to a safe depth. We went 45 meters under water and continued our journey at this depth. We reappeared shortly before the city of Chanak. When the tower came out of the water, the hatch opened quickly and the German war flag was shown so the batteries on the heights would recognize us. As soon as we were above water and noticed by the troops, they greeted us with roaring shouts. The cheering continued from height to height accompanying us continuously on the whole journey. We docked next to an old Turkish warship in the port of Chanak. Now we were safe because enemy fire did not reach so far.

The first thing after docking was we cleaned ourselves thoroughly. Washing at sea was a great luxury for submariners. The officers were no exception because existing water supplies had to be used sparingly. If you were allowed to wash yourself at sea, it was a great event and was officially announced beforehand with due importance. Captain Hersing once put this inevitable evil into apt words: Torpedo [boat] driving is a dog's life, but submarine driving is a pig's life. With that he was telling the whole truth about washing.[23]

---

[23] U21's War Diary continues the extraordinary events narrated in this chapter:

May 24 Quick dive, battery is almost full. Went to 20 m. Both motors F.K (half battery). Course 150°.

May 25 Went to 10 m. Course 160°. Course change after land bearing to SW.

345am Some warships abeam St.B.at anchor in the bay at Cape Kephalos (Imbros Island), surrounded by numerous merchant ships, transports and smaller ships.

4am French battleship "Suffren" comes up from aft. Abeam B.B. at Gabe-Tepe a single ship of the line (type Triumph) at anchor. A large number of ships ahead under land; Underneath clearly some large warships (1 tripod mast can be seen). "Suffren" passes coming up from astern, out of range. I decide, as I did with Askold, not to attack him either, in order to get to the English heavies first. The shot at the ship in motion is not very promising in the smooth weather either. Course 170°. The majority is not, as I expected from the news available, at Gaba-Tepe, but at Cape Tekeh, about 8 nm further south, so an attack on the heavies is no longer possible at dusk. With the perfectly smooth sea and the very clear weather prevailing here, the attack will be difficult in broad daylight, but on the other hand it promises success, since the boat has apparently not been noticed beforehand and the ships are anchored.

530am Passed 2 English destroyers (Arun-Kl.) At 4000 m. Course 140°. Surrounded by many smaller vessels, three battleships, (1 Agamemnon, 1 Triumph, 1 Majestic).

550am Went to 20 m. A hospital ship that is passed submerged turns hard towards the boat and drives away over the boat. Apparently the periscope has been noticed.

605am A destroyer (Basilisk-K1.) Is near the boat and drives over the boat.

620am Prepared to attack the first stationary battleship (Triumph type). This one has no nets. The "Agamemnon" has nets and is surrounded by many steamers.

625am Approached to within shooting distance. The ship of the line (apparently notified by the hospital ship), although I was traveling at 12 1/4 m and the periscope was used very carefully, noticed the boat and turned the bow towards the boat with its anchor motor. When the periscope is shown again, the ship opens fire at approx. 700 m from the forward turret. The impact is heard very loudly in the boat. Went to 20 m.
The destroyer, who had previously stopped by the boat, often drives over the boat and stays close to the boat so the attack has to be abandoned.

630am At 25 m with 340°. expired. The destroyer's screw noise gradually fades under land.

840am Went to 10 m. Under land a Majestic class ship of the line underway. Proceeded to attack. Behind the stern, a Queen class ship with medium speed passes by at a favorable range.

905am Rear shot at about 1000 m. When the periscope was extended to observe the shot, the ship dodged the torpedo by going astern, turned and opened fire on the periscope with the aft light guns. Apparently the ship saw the air bubble of the discharge in the smooth sea and was able to turn away in time. In the smooth water I have to forego shooting at targets on a bright day. On the other hand since the battery is insufficient to attack in the evening and then travel far enough away, I want to attempt an attack on the ship (Triumph type) that is still anchored under land near Gaba-Tepe. Further under land are 2 ships of the line of the Majestic class, surrounded by numerous auxiliary ships. The ships lying at Cape Tekeh are apparently anchored eastward for battle.
The ship of the line (Triumph type) has torpedo nets out. Near him there is balloon ship with captive balloon. The ship of the line is orbited by a destroyer at high speed, which occasionally forces you to go to about 20 m during the approach.

1120am The destroyer drives away over the boat at 15 m.

1123am Bow shot with N.S. Shot distance 300 m. A double detonation is heard, a quiet metallic thud, immediately followed by a very loud crashing explosion. In the interests of our own safety, it is impossible to observe the effect of the shot, since the destroyer rushes over the boat at high speed again immediately after the shot. When the boat is already at 20 m, two projectile detonations are unmistakably heard nearby. They seem to have seen the tracks of the ejection bubble or the periscope; both could be seen in the perfectly smooth water.

The noise of the destroyer's screws is also in close proximity even when traveling at 20 m and despite changing course, so one has to do without a periscope.

Went to 25 m. It should be considered whether the light gray color of the boat can still be seen at 20 m due to the transparency of the water. Through the central periscope you can see at 25 m to the front lifting hook on a bright day. An obscuring coat of paint, perhaps blue as on the Austrian boats, should be applied on the semicircular side walls, the gray tops of which is still now visible.

140pm 330°.

220pm Went to 10 m. At Gaba-Tepe there is no longer a ship to be seen. The balloon ship was also gone. Numerous smaller vessels at the launch point. The ships to the south have also left the anchorage. Continue to 25 m.

12am I intend to go back to W to proceed north around Samotrace, use the time available to recharge the battery and then perhaps go to the berth of the "Askold" in the evening.

At night I want to go south again to be at the southern tip of Gallipoli at dusk.

May 26

1025am Repairs on the St. B. oil engine (exhaust valve hot) 3/5 knots speed. 1 Charged.

120 Repair finished. 7 knots speed. Course 90°

5pm F. T. erect. No signal.

May 27

245am Went to 10 m to determine position, since the sky was cloudy and the moon has set; it is too dark to be able to proceed at 10 m. Continued at 20 m. Course 160°.

320am Went to 10 m. 180°. No ship at anchor at Gaba-Tepe. South of Tekeh-Burnu a battleship among numerous transport, hospital and other auxiliary ships, circled by 6 destroyers. The transport ships are all empty; nothing to see of troops landings. Camps on land. The battleship was so closely surrounded by auxiliary ships that a shot approach with a favorable intersection angle seemed at first quite impossible.

415am Maneuvered to attack, first on a westerly course, then on an easterly course. Sea conditions for unseen approach somewhat more favorable than on the 28th (Sea strength 1-2).

538am Bow shot between two steamers at a range of approx. 6 - 700 m. Intersection angle about 120°. Detonation is heard. Boat trims down somewhat after the shot. The destroyers cruising nearby approach the boat and repeatedly pass over the boat.

Went to 20m.

615am Went to 10 m. The hull of the capsized ship can be seen above water between two steamers. A number of vessels, 8-10 destroyers and several fishing steamers are systematically searching the sea area between Imbros and Gallipoli.

In order not to reveal where the boat is withdrawing to recharge the battery, the journey continues at 20 m.

May 27

1am Went to 10 m. every two hours.

Several steamers in the bay at Cape Kephalo. A Queen class ship close under land in 10 1/2 fathoms of water, under the protection of three dense rows of steamers. Attack does not promise success, especially as destroyers and fish steamers follow the boat.

4pm The northernmost security line is apparently between Cape Suvla and the northern edge of Imbros. The line is already astern at 4 o'clock.

640pm Surfaced. Two guard vessels aft are still in sight, which are slowly disappearing. 330°. 9 knots speed. Charge. Daily consumption: 6000 amp/hour

May 29 Ship targets are nowhere to be seen at sea, and an attack on ships while underway does not promise success.

11am Went to 10 m. In the bay of Cape Kephalo, the Queen battleship, which was observed on the afternoon of May 27, at anchor behind several steamers. I will try to shoot through between the steamers.

1230pm Went to 10 m. Immediately went down to 20 m; Seaplane is just about to take off from between the steamers.

110pm Went to 10 m.

114pm Struck at 6 points to St.B. a low barrel buoy, approx. 50 m at the same time on port side an identical buoy, close to it a few Balkan. I realize I have a barrier right in front of the stem, immediately go to 20 m and turn hard to St.B. from. Lines falling down onto the upper deck are observed through the central periscope, on one line a small, unrecognizable body, which, according to U 32's experience in the Channel, is initially mistakenly identified as an explosive device. It was a lead weight of the net. Using hard starboard rudder and full astern on the starboard engine (port engine full speed ahead) the net tore and came free. The net gradually falls all the way down onto the boat. 2 large floating hawsers visible, awash in the bay around the stem, aft and above. Moved slowly at 45° for 25 m, apparently the boat working in the net has not been noticed above. Pursuing screw noise is not heard. There is only the risk that the ends of the floating hawsers or even a barrel buoy will be dragged up and seen by a vessel happening to pass by.

230pm Down to 30 m. Screw noise passes close to the boat.

630pm Manage to free the boat from the net after an hour. The work was made easier by the fact that some of the net only had to be slipped back over the net deflector, it had-as a result of the deflectors - only hooked in a few places where it had to be cut. The torn off part of the net consisted of an approx. 100 m long hawsre filled with plant fibers, similar to raw cotton) and square wire meshes approx. 2 1/2 m in length, wire thickness in the meshes 8 mm in circumference. The torn off part of the net consisted of an approx. 100 m long Floating cable (a tube filled with plant fibers, similar to raw cotton) and square wire meshes approx. 2 1/2 m long, wire thickness in the meshes 8 mm in circumference. The meshes were spliced and soldered and covered with pieces of lead. The height of the net area was not more than 10 m.

	The boat was completely caught in the net above and below from the stem to the tower.
May 30	I intend to go south again tonight and stand by Gaba Tepe at dawn to look for warships. From Cap Teken I want to go on to Tenedos. Should no more targets be found, I intend to leave my position in front of the Dardanelles in order to first go to Orak Island according to orders. Since no news was received with F. T., I first want to find out about the situation in front of the Dardanelles at the news position. I would like to avoid staying longer than possible without news and with no prospect of further effectiveness, mainly because personnel efficiency forbids it. Even the mechanical equipment, although still fully ready for use, damage of a lighter and more serious kind (leaky seals and the likely leakage of the St.B. air cooler) is noticeable, which makes unadvisable to overstrain the mechanical capacity before clarifying it where I can count on the possibility of repair ...
June 1	
2pm	Orak Island on B. B. Came into Kishlebuku Bay. Rowboat ahead, the E.S. shows. Oberlt. z. S. Bieler on board.
240pm	Anchored with 45 m chain in 18 m water. Kaptl. v. Mohl on board. Received news of the situation from him. I intend to sail again the next morning, once again, coming from the north, to drive past Gaba Tepe and Tekeh Burnu and on the morning of 4.6.15. to enter the Dardanelles ... Nothing is taken on above the existing oil. There are still 33.6 t of fuel oil and 9.5 t of lubricating oil (approx. 2500 nm act. Radius) on board. I don't want to take oil on board beyond what is needed, as the boat is not able to dive during the transfer. Furthermore, since the transfer from the oil drums would have to be done by a submerged tank, I expose myself to the risk of leaving a trail of oil behind the boat during diving trips. Finally there remain one afternoon of necessary rest for the staff. From the lookout on the mountain, a fire is observed on the mountain top of Kos at nightfall. It is suspected that the sailor who was hit on Orak Island reported the boat.
June 4	Apart from torpedo boats, no warships were encountered. Steered into the Dardanelles. When turning, the boat turns into the current vortex at Cap Helles, not to port despite one of the engines reversing. It is turned to starboard in front of the current ....
11am	Strong detonations near the boat. Since at the same time screw noises, which stop temporarily, can be heard, it can be assumed a destroyer is chasing the boat with explosives. Detonations are sometimes very strong, close to the boat. Went to 30 m. The detonations continue. Down to 42 m. The European coast is headed for after presumably matching the entry current strength, at this depth not known.
1210pm	Went to 12 m. The detonations can only be heard very faintly in the U. T. handset.
1pm	Surfaced. The boat arrived at the barrier faster than expected due to the current at the deep depth. E.S. exchanged with the shore stations astern; in front of the entrance the French cruiser "Bruix" and several destroyers, which are close to land. The detonations heard were apparently projectile impacts by the Turkish batteries, which returned fire.
120pm	Oberlt. z. S. Prince Reuss on board. Have been piloted through the barriers.
June 5	Entered the Golden Horn
5am	signed: Hersing.

At 7 o'clock in the afternoon we continued our journey to Constantinople. It should not be forgotten we were received with the greatest admiration in Chanak as well. Both our long voyage and the sinking of two English battleships, which has since become known, made the deepest impression on the city's inhabitants and occupants. It was therefore expected to treat the heroes of such acts with the warmth and distinction inherent in the Orient.

In the expectation of soon reaching the legendary city of Constantinople, a decidedly festive mood took control of the crew. It was much livelier than before because a number of truly serious dangers had already been successfully endured. We reached Constantinople on Saturday, June 5th, at 5 a.m. As we approached, crowds lining the bank to greet us broke out in roaring cheers. We were waved at with arms and legs, and with large and small towels in all colors. The wonderful image of the city and enthusiastic calls of adulation increased our already festive mood. We did not expect such a reception. At 7 o'clock we docked at the port shipyard. Representatives of the naval and military authorities appeared to greet us.

Before landing, we had been able to cheer our comrades from the High Seas Fleet as we drove past the cruiser Breslau. The crew gave us a broad hurrah, which we responded vigorously. There was no doubt about it, the warmth of the greeting and loudly expressed enthusiasm was genuine and it filled us with genuine pride. But our successes were unique and unparalleled. We had every reason to be happy about the splendid reception. The focus of the adulation was, of course, our commander Lieutenant Hersing. He was surrounded on all sides, and everyone tried to congratulate him as soon as possible and with a firm handshake. We were all happy he was honored so effusively, we were proud of him and, with sincere admiration, granted him the abundance of awards.

After the very warm welcome, we hoped to be able to take a good rest from the hardships of the past few weeks. How great was our disappointment when the chief engineer informed us the boat had to be overhauled as quickly as possible and the necessary repair work had to be carried out by us. On Admiralty orders, U 21 should be fully operational again in the shortest possible time. Already at 9 a.m. there was a line-up for duty assignments. It was incomprehensible to us how the Admiralty could make such demands regardless of the men's condition, since it must have known regarding the enormous efforts of the crew. And we had been successful too. Overtired and reluctant to go to work, we went to work. Everyone was bitter the well-groomed people at the green table made such a request after we had done our duty day and night in the most dangerous post for weeks. On board we were of one voice about this. Of course, we still carried out the necessary work with the greatest care. We knew how necessary they were, and the slightest negligence could have the most pernicious consequences. Even our extremely depressed mood could not change this conscientiousness. After work we received our wages and were then allowed to go ashore. For the first time we received Turkish money. Of course, we did not get in the right mood that evening, as we had actually longed for.

With undiminished enthusiasm we continued the repair work over the next few days. The great heat was uncomfortable and almost unbearable. After a few days I was so frail I had to call in sick. The senior staff doctor determined it was genuine overexertion and said a fortnightly recovery would help best. But before he could order this, the chief engineer had to be consulted as to whether he could give me leave for that long. However, the latter stated it was not possible for him to do without one of his men for so long. Nothing came of the fortnight's rest. But I was released from work for three days.

The shipyard layover was much longer than initially assumed. A thorough inspection revealed defects requiring thorough repair. The layover time was extended to weeks, so we had enough time to have a good look at Constantinople. For myself, I made extensive use of this opportunity. But other comrades were also keenly interested in touring. We particularly liked the magnificent mosque buildings. We also visited inside some of them. In the Turkish quarter there were only old, dilapidated houses and the streets were unprecedentedly dirty. The Europeans' quarter, on the other hand, was just like a European city: beautiful wide streets, electric trams, large department stores, hotels and an abundance of entertainment venues. The number of dubious women loitering in the city was noticeably large. We had been advised on board of the dangers of consorting with them. We were told we had to be particularly careful, as we had to consider keeping intact the entire crew of the submarine, which had previously stood together in an exemplary manner. It would be impossible to find a replacement here if someone had to be depart because of a contagious disease. On this account the interests of our comrades required the greatest circumspection. This warning from our captain was heeded by everyone. Nobody had to be left behind.

Our stay in Constantinople now become more pleasant than we had every reason to guess on the day of our arrival. We could not ascertain whether and how much we had to thank our captain for the extended stay. In any case, we happily agreed.

The technical facilities of the shipyard were regarded as primitive by our German standards. It was only through the presence of many German shipyard workers that it was even possible to make real repairs.[24]

On July 1st, the entire U 21 crew was awarded for outstanding performance the Iron Cross II Class, the Turkish Crescent and the Turkish Leakat Medal by Admiral Souchon.[25] The admiral thanked us with appreciative words for the successes so far and asked us to continue serving the fatherland in the same excellent manner in the future. One day we also received a request from Enver Pasha[26]. This visit came as a surprise and without prior notice, because no

[24] U 21 would be bedeviled by inconsistent Diesel engine function during its tour in the Mediterranean.

[25] Admiral Souchon was the commander of all German naval forces in the Mediterranean at this time.

[26] Turkish Minister of War and one of the triumvirate who were defacto rulers of Turkey during the Great War.

preparations had been made for the reception. Captain Hersing himself led the guest through the whole boat in which we were all at work. Enver Pasha showed great interest in the equipment and conversed in the friendliest manner with a number of the crew.

Just as everything comes to an end, so did our stay in Constantinople. On Saturday, July 3rd, 1915, our preparations for sailing were finished. Again an unmistakable crowd gathered. When we cast off, enthusiastic calls rang out again from many thousands of throats. Some women even called out the names of individual members of our crew, proof that various very intimate friendships had been made, which suddenly came to an end in soldierly fashion. In honor of those affected, it should be mentioned that saying goodbye was not as easy for them as it outwardly appeared. In a sad mood within them arose the question: Who knows whether we will see each other again. But we didn't have time for long deliberations. Soon the two bridges closed behind us and we drove around the Golden Horn back to the open sea. In the morning at four o'clock we docked again in Chanak. On Sunday, July 4th, we were back in front of the Dardanelles to search for more prey. At 9:30 am the order to dive suddenly came because the French mail steamer "Carthage" was sighted. Our leaders knew it was now being used for troop transporting. It was now a matter of maneuvering with particular caution so the rare prey would not escape us. The danger of counterattack was not as great as with the heavily armed and often escorted warships. But the effect must have been extraordinary, and this explains why we cruised back and forth for several hours in order to get an exact shot. After more than four hours the time had come. The torpedo left the tube at 1:47 p.m. and the explosion took place at 1:48 p.m. The steamer immediately lay on its side and sank in a few minutes. We had once again destroyed 12,000 tons of shipping space. 3500 people are said to have sunk with the steamer.[27]

[27] The 5600 ton Carthage was commanded by Captain François Antoine Vecchioli (1863-1956) commanding. Extract of Carthage's captain report follows:

For the first time we were able to sink a large troop transport steamer. As proud as we were of it, the thought of the horrific scenes that played out upstairs depressed us. The panic that ensues after a torpedo explodes on a struck ship is indescribable. Especially if it is a troop transport steamer. The tween decks are crammed with soldiers and their luggage. The torpedo tears a huge hole in the ship's side, into which the seas flow; everyone rushes to the exits and the rescue facilities. The exits are blocked and personnel trapped below decks are hopelessly lost. Those on deck are not doing much better. Once struck the vessel sinks so quickly there is no longer any opportunity to unleash the lifeboats and launch them into the water; but then most of them are also unable to swim and those who can are dragged down by heavy clothing. Terrible fighting occurs in the water. One clings to another, whole tangles of those who wrestle with the waves disappear beneath the sea. Even when help comes, rescuing those fighting in the water becomes so difficult that most are lost. Everyone wants to get into the lifeboat first, one tries to displace the other, hideous scenes take place in such a fight.

On June 30, we started unloading ammunition, the rest of the load to be unloaded at Mudros. To save water and coal, we have kept only four boilers in operation. From June 30 to July 4 the unloading was carried out with some interruptions because of the changes of anchorage made necessary to avoid the projectiles fired by the enemy batteries on the auxiliary transports at anchor.

On July 4 at around 1:48 pm, the cry "A submarine! Was yelled by men onboard. Finding myself on the deck, I ran aft and saw the wake of a torpedo which was only 100 meters from the side. It struck the port side of Carthage a little behind the engines at 1:50 p.m. The ship, shaken by a violent shock, instantly began to sink by the stern with rapidity, then raised its bow almost vertically, immediately fell and sank. It was 1:54 p.m.

Five barges which were moored alongside to receive ammunition, as well as the tug Phocen, from Algiers, which was on towing duty, were used for the rescue. The operation was carried out in good order and without panic. The first lieutenant and mate led the forwards operations and I led the aft. Belts and lifebuoys were quickly thrown into the sea. Boats from the ships Bretagne (hospital ship), Sainte Helene and an English ship, as well as the motor boats of the Port Authority were immediately brought to our help. When the rescue was finished and the scene of the disaster explored to ensure that there was no one in danger, all the men of the crew, the Goliath mission and 5 passenger soldiers were gathered on Tireless. We made a roll call and regret to note the absence of six men from the crew.

In general, the troop-transport steamers only had to travel a short distance. They brought the soldiers gathered on the islands of Lemnos and Imbros to mainland Asia. Because of the short distance, the transport steamers were fully loaded to the extreme limit in order to bring as many soldiers as possible in one trip to the theater of war. When such an overcrowded steamer was sunk, it was an unpleasant and palpable loss for the enemy. The sinking of the steamer "Carthage" was also a heavy blow for our opponents. The consequence of this act was it became a little too uncomfortable for the English. During our further cruises we could not find any more vessels in this area.

On Tuesday, July 6th, we cruised again in front of Saros Bay. Then we were noticed by a guard ship, which immediately approached us. It came to life in the boat. Alarm. Quick dive. In a few minutes we were down to 25 meters. Now we felt secure and believed we were protected from sight. But water in the Mediterranean in general and especially in front of the Dardanelles is clear and allows quite a deep view. A submarine had to descend to considerable depths if it did not want to be seen. We felt safe from all dangers at a depth of 25 meters. Above us we heard the steamer we saw driving by, and we clearly heard the engines pounding. But suddenly we were torn from our rest. All of a sudden there was a tremendous explosion, our boat trembled in all parts, lights went out, the lens on the engine telegraph in the tower shattered. The crash and clatter in the boat seemed so unsettling everyone thought the last hour had struck. I was just standing at my diesel station. The iron plates under my feet pushed in and out with such force that I was knocked over. The boat was tossed back and forth like a ball due to the enormous water pressure. From the tower came the command from Captain Hersing: "Check everything to see if the boat has a leak anywhere!" Everyone who could now searched but the boat was watertight everywhere. We turned around at once, and Captain Hersing, who kept the greatest calm, let the periscope up to see where the steamer was. To everyone's surprise, Hersing announced: "The steamer has sunk; you can only see some debris floating around." Of course we asked ourselves: what happened there? We didn't torpedo! From the presently developing guesswork it was believed we had got into a mine field and torn off a mine with our mine fender. In any case, when the mine came up, the steamer, which was just behind us, must have thought we were coming up. Without further ado in this erroneous assumption, he tried to ram us, but as a result of his mistake he drove full steam onto the mine instead of us, which naturally exploded and resulted in the steamer's sinking. We were lucky our boat had run so deep. As a result, the effect of the explosion was greatly dampened. Nevertheless, after surfacing, we found the strong water pressure had left a deep dent on the upper deck of U 21.

From the increased number of vessels looking for us we soon recognized our enemies had organized a large scale hunt. The Dardanelles were also practically contaminated with mines. Many destroyers patrolled the waters always looking for us. Apparently under any circumstances they wanted to kill the hated German submarine.[28] For several days we did not relax due to our persecutors. We had to be on our guard all the time so as not to be suddenly surprised. While we could venture to the surface during the day to identify an approaching danger and ventilate the boat's interior, Captain Hersing preferred to stay underwater at night, as this meant the greatest possible safety for us and the boat. The crew found the nocturnal underwater stay more pleasant because at least there was a certain calm. It only had the disadvantage that part of the crew had to remain at diving stations. The off duty watch was allowed to lie down in any corner of the duty station.

I spent many nights in the "rest period" on the passageway slabs because there was no other place to lie down. However, this spot had a major disadvantage. Anyone who wanted to go to a private spot had to pass me. It now became real sport to give me a gentle kick when anyone passed. As the passage was so narrow I almost

[28] In his report, Hersing writes of the enemy reaction: "The enemy maintains such surveillance here by means of destroyers and trawlers which use all the means of combating submarines already employed in the Channel that the chances of success are quite small. Added to this is the fact the calm sea and water clarity make it easier for the adversary to find submarines."

completely filled it when lying down, there could be no question of real rest. But what could you do? You were happy when you were under water. Cruising surfaced was extremely stressful at night because the boat could be overrun by one of the nimble destroyers at any moment. Since every warship was dimmed during night, it could only be seen at very short distances in the dark.

We didn't experience anything exceptional in those stressful days of back and forth cruising above and below water. The sinking of the aforementioned steamer was to remain the only event. Therefore, after a few days, we tried again to proceed safely through the Dardanelles Strait to our home port of Constantinople. Captain Hersing waged on luck for the breakthrough. It went smoothly and without incident, so that we arrived at Chanak on Saturday, July 10th, 1915. Then we continued through the Sea of Marmara to Constantinople. We all breathed a sigh of relief and were happy to have solid ground under our feet again. The events of the last trip were of course discussed extensively, with a lot of laughter. All efforts and inconveniences were quickly forgotten. On Sunday we reached Constantinople, again warmly received. U 21 was berthed in the shipyard from July 11th to August 29th. Previous relationships on land were gladly re-established. The opportunity now presented had to be fully exploited after the privations previously experienced. An onboard party organized by the crew turned out to be particularly lovely. The musical part was taken on in a friendly manner by a Turkish band that alternated between German and Turkish music pieces.

The stay this time was much more pleasant than the first. Unfortunately, the beautiful days passed too quickly.

On Sunday, August 29th, our boat was ready to sail again and we waited for things to come. In the middle of our work, we received the order from the Admiralty to leave immediately, as there were English submarines in the Sea of Maramara. One of them had run aground and was not able to move.[29] We left at 4 p.m. At 5 o'clock we found ourselves below Stefano. Suddenly there was a flash on land, and to our greatest amazement, a missile hit the water less than three feet from our boat. Immediately afterwards a second one whizzed right over the tower. No doubt about it: we are being fired at from Turkish batteries. The signal pistol was brought up on deck as quickly as possible, and before the shore battery could zero in, a signal star was fired, whereupon it stopped firing. It had all happened so suddenly and unexpectedly that only now did we really understand what was going on. I was just off watch and was on deck with a few comrades. Captain Hersing was standing on the conning tower. When the first shot struck so close to us, he immediately shouted: Everyone get down! But in the confusion everyone got muddled up. Only when the second shot got over us did we drop on deck. If a sailor hadn't had the signal pistol so quickly at hand, we would have been shot down by our own batteries. But a second, even greater danger threatened us at the same time from a Turkish torpedo boat driving shortly afterwards. On this boat, too, the serious intention had been to kill us. It immediately headed for us after the star signal was fired,. When it drove close to us, its commander pointed extremely excitedly at the gun crew and said, turning to Hersing: "You see all the men here are still standing at the guns. I was just about to give the command to fire when, for the first time, I saw the star signal as a recognition signal. A few seconds more and we would have done it." Hersing, too, was angry about the incident. He said that it is not enough to face your enemy on the open sea and fight with them for life and death, then you are also exposed to the danger of being shot at by your own batteries. This incident, which almost was disastrous for U 21 and its crew, was due to the relevant authorities forgetting to notify the batteries and the other authorities in question of our departure.

[29] There is no record of an English submarine aground in August 1915 in the Sea of Marmara. Possible corruption of prior beaching of an English submarine in April, a French one in July, and a recent landing party operation on August 21st by an active submarine.

Since we didn't notice anything in the slightest about the reported English submarines, we didn't stay long, but set out on a long-distance journey. Now it started again, searching day and night for prey to wreak havoc and humans to kill. Sometimes in the seclusion of the submarine the thought occurred to you: Has Mankind actually gone mad? But what was the use of thinking. We had to do our duty.

U 21 stayed for a few days in the Gulf of Saros, where we had gained such rich prey. But this time we searched in vain. So we continued on. On Monday, September 5th, we were in the Gulf of Salonika. Sacred Mt. OIympus lay in front of us. In the evening we made preparations to lay on the seabed overnight: The boat is thoroughly ventilated again, a favorable location of about 45 meters is sought, and then it goes to rest at night in the depths. After a few minutes we feel the first contact with the seabed, and the boat lies down like a tired giant animal. The motors are turned off and everyone steps away from their diving stations. The watch is set, two men at a time, to go through the boat from time to time. The officers sit in the mess for a while with a cool drink. Some comrades from the crew sat down to play cards, others are standing around the gramophone, listening to German marches and songs, and some sit quietly in the corners, leafing through their letters, looking through photographs of their relatives and looking at them again and again. It gets quiet around midnight. The changing of the watch is only awoken every two hours. Otherwise there is dead silence in the boat. However, the air gets worse and worse and breathing becomes more difficult.[30]

My turn comes to watch too. I am gently nudged and immediately fully awake. I check the clock. There are ten minutes before I have to relieve the watch. With my eyes open I lie still and cannot banish the thought that we are actually buried alive here. A single failure of

[30] The two men on duty are to ensure fresh oxygen is provided as needed and no noxious gases are generated, (chlorine gas is created when seawater comes into contact with sea water) while the remaining crew slumbers.

the diving devices and the boat becomes a coffin for thirty-one men from which there is no escape.

Awaken at 7 a.m. Stations are manned immediately and the order to surface is given. The chief engineer, the captain's right-hand man, stands in the control room and gives orders to his technicians. The boat slowly rises from its resting place and soon we are on the surface. As soon as the tower is above water, the hatches are immediately pulled open to let fresh air into the boat. Their beneficial influx is welcomed by all. One by one, the men off watch go up on deck, armed with a cigarette. That is a treat after so many hours! Soon the cook will have the coffee ready, and we will go back into the boat to satisfy our hunger.

For the next few days we cruised the blue waters of the Aegean Sea. First along the Greek coast. We spotted a French steamer with two funnels in the distance. Unfortunately we couldn't reach him. Then a course was taken across to Lemnos. We made an approach on the four funnel steamer "Mauretania". The torpedo was properly shot, we even notice the impact on the ship, but to our utter disappointment it did not explode.[31] A second attack was no longer possible.

For the next few days we were busy looking for new prey. However, our enemies had become very careful. There was no longer any opportunity to inflict tangible losses on them. We spent most of the nights under water.

During this time the English had begun taking more decisive defensive measures. We learned of this when we were in the Gulf of Saros on September 13th. Here the news was conveyed to us that the enemy had covered the whole approach to the Dardanelles with mines. So we waited the next few days in the Gulf of Saros for further orders. After a few days orders came for us to longer call at

[31] On September 9th, 1915 U 21 missed Mauretaina with a torpedo, 9 miles North of Strati Islands. Torpedo was fired at 1400 meters, Mauretania was departing Mudros after transporting troops there on September 3rd.

Constantinople because of the great danger of mines. Pola in Austria would become our new home port. The commander's orderly who brought this order also handed us the last mail. As always, there was considerable happiness in receiving a sign of life from home. It always took several weeks before any news got into our hands. However, our joy was clouded by the fact we had left our belongings in Constantinople assuming we would repeatedly return there. We were thoroughly mistaken about that. The clothes and souvenirs that we had at the Golden Horn were later forwarded to us, but a great deal was missing.

The journey to Pola went through the Aegean Sea past many beautiful islands. At a brisk pace we reached the island of Crete in a short time and, turning west, drove around the southern peaks of Greece. We entered the Adriatic Sea again through the Strait of Otranto. Here we first set course for the familiar port of Cattaro. After a short stay, the journey continued to Pola, where the U 21 received a longer shipyard refit time, which lasted until January 1916.

In October I finally got my long-awaited home leave. Home leave! Whose heart didn't beat faster? For eighteen full days I would be free from this arduous boat service, for eighteen full days I should belong to myself again rest from the sometimes almost superhuman exertions. Filled with the images of a carefree leave time, I said goodbye to my comrades and superiors. On the express train; the journey first went via Ljubljana to Vienna. The train carried us through wonderful areas, past beautiful cities and high mountains, through areas one had never thought one could ever see. In Vienna, the express train to Berlin was taken. From here it was immediately on to Cologne, from there a short drive, and one autumn morning at 5 a.m. I finally stood full of excitement at my parents' front door. Everyone was still in the deepest sleep. I knocked on the door, after a while a window opened and my father's voice asked: Who is it? But he recognized me at a glance, and his loud joy immediately woke the whole house up. After the initial boisterous greeting it was time to tell the story, for everyone was proud that one of the family was part of the crew of the famous U 21. Of course, word of my arrival in town got around very quickly. Friends and acquaintances came to admire the submarine hero. Everyone wanted to hear in detail from my mouth the wonderful things about which the newspapers had only reported in short articles. I received invitation after invitation because nobody wanted to miss the opportunity to host a real submariner as a guest. Everywhere my appearance gave cause to celebrate a little party. There was still enough of everything, you didn't need to limit yourself and everyone was absolutely convinced of our victory. So the days passed like real vacation days, just as I had dreamed of them. Unfortunately too fast. Never before had I found it so difficult to say goodbye. I thought with horror of the coming submarine voyages, of the exertions and strains that went with them. But what was the use of all this? Duty called.

After a warm farewell, I started my return journey to Pola. Although the trip again offered many attractions, I couldn't get in the mood; my thoughts were still at home. But the closer I got to Pola, the more my thoughts were of those comrades with whom one had shared joys and sorrows, and that sense of togetherness that no other outfit had. Duty began after hardly on board and reported back from leave. I quickly settled in again and found myself on the old track. The boat was painstakingly overhauled. This time it lasted longer because it was not just a matter of checking the machinery, but also utilizing all the advances that had been made in the field of submarine warfare since our first departure in April. That is why we could celebrate Christmas happily in this country. In addition to many small innovations, our boat had a new quick ventilation system, which should enable faster diving. Lengthy practice cruises and diving tests were made to try out the various changes,.

On January 22nd, 1916, U 21 set out to sea again for a major cruise. Our destination was no longer Constantinople, we were given the task of navigating the entire area of the Mediterranean. After only a week we were cruising back and forth off the coast of North Africa, always looking closely for prey. We finally got lucky. The 8000-ton steamer "Belle de France" was sighted and attacked without warning.[32] The fired torpedo was a direct hit. The crew just had time to send out distress calls and lower their lifeboats in order to escape as quickly as possible from the steamer. But the sea was very rough, sufficiently for lifeboats to capsize and the crews attempt a swim to our boat. Most of them sank before our eyes; we were only able to take 13 men on deck. They were Indians. We learned from them the steamer had been commanded by two white men. On looking for them we found them sailing a good distance away, without bothering about the fate of their crew. From the rescued we further learned the two had lowered their lifeboat alone on the opposite side of the steamer.[33] The Indians we took in complained pitifully because they thought we would make short work of them.

---

[32] The unarmed British flagged 3876 ton Belle de France was carrying grain from Karachi to Algiers. Captain James Cargill commanding. Attack took place 126 miles N/NW of Alexandria on February 1st 1916. Hersing attacked without warning on this occasion despite Mediterranean submarines operating under Prize Rules. Possibly due to U 21's engines operating erratically limiting surface speed to 12 knots and the ability to pursue escaping vessels as recently some vessels had escaped due to their higher speed.

[33] The British Captain's reported differently. A summary of his report follows: At 2 p.m. on Feb. 1, when, without any warning, the ship was struck by a torpedo on the starboard side in the way of the cross-bunker holds. She listed heavily to starboard at once, and made much water, the hatches from No. 2 and the cross-bunker holds being burst open.

All hands were promptly called to the boat-stations, where the boats had already been swung out in case of attack. No. 1 lifeboat on the starboard side was then found to be broken by the concussion, and useless. The ship was listing more and more heavily, and all hands were ordered to the port-side boats, two of which were lowered by the master and officers. The lascar crews were ordered to keep close alongside the ship, but they became panic-stricken and cast off from the ship without orders. The master and three other officers jumped into the boats from the deck. The chief officer, who was standing by the falls, and the chief engineer, who was stopping the engines, were left on board as the boats drifted away.

The chief officer dived overboard, and was picked up by No. 3 boat. The chief engineer, being unable to swim, remained on board till, as the vessel righted herself. he succeeded in getting into No. 2 starboard boat, which was partly lowered. After about half an hour he was picked up by No. 3 boat.

No. 4 boat, in the meantime, had capsized. Some of the crew were swimming, and others were clinging to her bottom. The submarine rose to the surface, came alongside, and picked up these men. No. 3 boat was then called alongside the submarine by the officer in command, and was ordered to stand by. The officer of the submarine took his revolver and threatened to shoot both crews if they came nearer.

We were still busy with their questioning when suddenly a cloud of smoke was reported from the tower, which was approaching us quickly. Apparently it was a destroyer who, in response to the distress calls from the steamer, which has since sunk, was approaching at full speed to the scene of the incident. Alarm! Quick dive! Everyone rushed to their diving station, the hatches were closed. Of course we couldn't take the Indians on board. They had to be left on deck for our own safety. The boat quickly began to dive. When the Indians saw what was happening, there was a terrible panic. They tried to force the hatches open to find shelter in the boat. But they did not succeed because we held the hand wheels from inside. We sank deeper and deeper and the men were swept away by the waves. We were already 15 meters deep when we noticed, to our great astonishment, that one of the Indians was still holding onto the tower railing. But while diving deeper he suddenly let go and went up quickly. We could no longer see what had become of the men. We assumed that at least some of them were taken up by the destroyer.[34]

In the next few days, our trips continued along the African coast. We paid special attention to the area off Tripoli. But there were no significant events. In vain did we search for prey. The otherwise busy lines were without traffic. Our captain therefore decided to pay a visit to the eastern part of the Mediterranean. Since there hadn't been a submarine in this area before, we could assume that our enemies would hide less fearfully here. We first set course to the cities of the Holy Land. We cruised back and forth in front of Jaffa, Beirut and other smaller seaside towns until we were finally lucky

---

At this moment, four trawlers were seen on the horizon, and the submarine, sublimely oblivious of the shivering men it had just hauled on to its deck, dived with the whole bunch of them still standing there, and left them to flounder to the surface as best they could. Some of them were saved by No. 2 boat, but nineteen were drowned, a good many being sucked down by the diving submarine. A delay of a very few seconds, of course, would have made it possible to save them all.

[34] British authorities regarded the incident as a war crime incorrectly believing the submarine had deliberately dived with survivors on deck in order to kill them. U 33 commanded by Konrad Gansser was misidentified as having committed the sinking. The case was investigated post-war as Naval Case No.21 and eventually dropped as it was found the submariners rescued survivors from an overturned boat and called to other lifeboats to close and take off those on the submarine.

again. On February 8, in the pale twilight of the beginning of the day, the keen eyes of the lookout saw the outline of a large steamer. Immediately we dived and carefully approached. As we came closer, we recognized the enemy: it was the French cruiser "Admiral Charnier". Finally once again a ship from the "Majestic" and "Triumph" classes. Wild joy seized us. We knew our experienced captain would take the greatest care to keep the enemy in front of the muzzle, i.e. to place a torpedo in front of it. As always, it was stalked with the greatest caution, every command was carried out with the utmost speed and meticulousness. There - suddenly the command: "For the shot!" and shortly afterwards the last command: "Go!" We waited with the utmost attention for the impact. A minute passed, a second stretched out into an eternity, and there was still no roar indicating the destruction of the enemy ship. We already doubted the success of our otherwise almost infallible torpedo when the impact was finally heard after almost two and a half minutes.

The effect of the shot was of a horror that exceeded anything previously experienced. In a few minutes the ship was simply gone, dragging the entire crew into the vortex. As we later learned, only one sailor was reported to have been rescued.[35]

The terrible tragedy of the sinking of the "Admiral Charnier" swept enemy ships from the sea. All sought shelter in ports. Hence, our eager search for new prey in the period immediately afterwards was unsuccessful. We couldn't find any targets. Until February 12th, we cruised in the area south of Asia Minor. On that day, after sailing the Mediterranean for three weeks, we had to take course to Pola

---

[35] Quartermaster Joseph-Marie Cariou. Recently assigned to the ship, onboard for 15 days prior to the sinking. His testimony: "On Tuesday morning February 8, we were near Beirut, about 10 to 15 miles offshore. I was on the aft deck. The weather was fair enough, not very cold, very little wind, ... Having heard a thud with a strong shaking of the ship, we immediately had the impression that it was a torpedo, because the night before, while setting sail from the Isle of Rouad, a submarine was reported. The boat nosed down a little, and capsized almost immediately. It was then that I wanted to take off my jacket. I didn't have time, I was thrown onto the guardrails and I sank with it. When I came back to the surface, I found myself next to a small piece of wreckage, a demolished hen cage , and I stayed on it for about an hour...."

again to fill up supplies and provisions and give the boat the usual overhaul. On the way we were surprised by a strong storm for the first time. Like a nutshell, the U 21 was driven back and forth by the rising waves. Huge flash floods swept over us and completely buried the boat under them. The watch post on the tower was literally flooded by masses of water, so that their life was in danger. The seas even entered through the tower hatch into control center. As a result of being thrown back and forth it was impossible to stand on one's legs in the boat. Was it any wonder when a certain agitation took hold of the crew? We had already been through a lot, including some choppy seas, but such a storm had not yet been experienced. Fearful doubts arose: can U 21 pass such a test of strength! Will it be able to hold up against the ever-increasing waves or will it be crushed by the force of the falling water mass? How easily steerageway could be destroyed by the raging flooding and the crew left to destruction. The seas became increasingly agitated; the storm drove the waves higher and faster. Flash flood followed flash flood. In the end, staying afloat became a serious hazard. The only salvation now was to attempt a dive. Hersing issued the appropriate orders. But even when diving we were still lifted up and down by the waves, tossed back and forth. Even at a depth of 30 meters we still felt the influence of the angry sea. Only at 40 meters do we have full control over the boat again and could use the rudder as desired. The boat sailed at a slower speed and proceeded very carefully since it was assumed the boat suffered some kind of damage in the unheard of storm. But to our delight we were able to find out very soon that U 21 passed the tough test of strength with flying colors. Its seaworthiness hadn't suffered in the least. When we surfaced the next day, the sea was calm again and we were able to continue surfaced our journey to Pola. We entered the port without any incident worth mentioning.

I had had a painful ear condition during the last few weeks of this trip. I had this earlier on U 14 during a pressure test. The ear drum was torn and the slightest pressure in the boat caused severe pain. Probably as a result of the terrible storm, I was now constantly plagued by pain, even with normal air pressure, such that I was compelled to call in sick. Our assistant doctor sent me to an Austrian senior staff doctor. After an in-depth investigation, he reported me as unfit for on-board service in the submarine service. I reported this result and received my assignment to Kiel the next day. For the time being, my service on a submarine and as a participant in Mediterranean submarine cruises was over. I did not expect this elimination to come so quickly. This sudden change was not entirely right with me. You would have been a bad comrade if you could easily have parted with your comrades in arms, with whom you were most intimately connected and with so many common experiences.

When I arrived in Kiel, I immediately reported to the senior doctor in the submarine department. I handed him the referral papers I received in Pola, the examination results and the accompanying letter from our on-board doctor. After the senior doctor had read through the papers, he looked me up and down and from bottom to top. I thought he would now start the investigation. Instead he said: "Well yes, this is typical, you got enough medals and decorations, and now of course are no longer interested in the submarine service. But you can rely on it; tomorrow I'll go myself to the chief medical officer with you." Only now did he undertake a brief examination, or at least do something like that, after which he scribbled the result on a piece of paper. Then I could go. The next day I had to go to the hospital to see the chief staff doctor. He did the examination, however, much more thoroughly than the senior physician and, moreover, in the friendliest way. The result was the presence of a severe scar and significant hearing impairment, such that I was eventually declared unfit for service. So, I thought, now the war is over for me. I will do the rest of my service with the submarine department in garrison. Now that I had come here, I no longer had any desire to serve on board a submarine again.

After a few days, I received an easy but very interesting duty at the personnel office. Here I was entrusted with compiling lists of submarine crews. We found out first here about all submarine losses. The head of this office was a chief machinist with an engineering license, a very friendly man who had known me since my time in peacetime service. In a few weeks I became well acquainted with life in the barracks. It was like peacetime. But but! Gradually it became noticeable Germany was cut off from all food supplies. Dissatisfaction with the increasingly poor and worsening diet grew slowly but steadily, and at the same time there was diminishing hope of a successful outcome to the war.

In my department in particular, one could understand the course of events. Because here attention was focused on losses in the submarine weapon so important for our ultimate success. In public, these losses were less noticed than successes in the land war, while we knew exactly what they meant. We could not hope for a resounding success in the submarine war if it was not possible to quickly improve this weapon through technical advances.

IV. To America in a Merchant Submarine

In this mood of doubt came hope with the news a German submarine, U-Deutschland, under the leadership of Captain König[36], had arrived in America as a trading vessel. Once again Germany had shown the world what great technical deeds it is capable of. Amazement and admiration on the one hand, envy, resentment and hatred on the other. England in particular foamed with rage. After all, U-Deutschland had shown the entire world that English blockade measures were inadequate. A German ship drove through the huge blockade as if there were no obstacles; valuable cargo was transported to America and returned to Germany, the most important raw materials for continuing the war could be brought in against the will of the enemy. A major achievement of the first order. The use of a submarine as a peaceful cargo ship had not yet been thought of. All of Germany was now waiting for the first return of its distinguished vessel, and great preparations were made for an appreciative reception. The enthusiasm was enormous when U-Deutschland successfully and safely landed, fully laden with valuable freight. Now U-Deutschland was in the Germania shipyard making ready for its second trip to America.

The presence of the U-Deutschland brought a change in my sedate garrison life. In September 1916 I met the chief engineer of U-Deutschland, Mr. Kleese[37], whom I knew well in the hallways of the submarine department. Mr. Klees knew I had been part of U 21's crew. After greeting me, he asked me what I was doing here. I replied I was unfit for on-board duty due to an ear problem and was now doing office work. Mr. Klees said: "This won't be as bad. I need a capable, experienced submariner. You might be the right man I'm

[36] Paul König,(1867-1933) had earlier captained merchant submarine U-Deutschland in a June-August 1916 voyage to the USA.

[37] Helmut Klees was Chief Engineer of U-Deutschland.

looking for." In the further conversation, Mr. Klees explained to me I wouldn't ride as a stoker but as an assistant engineer. At the same time he guaranteed me a nice salary. I was very enthusiastic about the tempting offer. But I had to point out my current superior would hardly let me go so easily. But Mr. Klees said he would take care of that. The main thing is I agree to it. Then I went back to the office. After a short while, Mr. Klees followed me in and greeted his former comrade, the chief machinist who was my superivisor. The conversation immediately turned to the needed man, and of course I listened in the greatest suspense. Mr. Klees declared he absolutely had to have an experienced man, to which Mr. Mehlitz replied he only had personnel available who were unfit for on-board service. Then Mr. Klees asked whether I was out of the question. "Out of the question," explained Herr Mehlitz, "I'm not going to hand him over; he has only just now been properly trained. In addition, he is also unfit for on-board service;" Mr. Klees should see whether he could find somebody in the shipyard who was suitable for his purposes. Mr. Klee was actually put off and left. Hope which had risen up in me faded into nothing. I would have loved to have served on the merchant boat. When Herr Klees left the chief engineer asked me if I would like to go on the merchant boat. Of course I said yes.

Herr Klees must have had no luck at the shipyard because the next day, very early in the morning, he came back to Mr. Mehlitz and now declared in the most determined tone that he absolutely had to have me. Mr. Mehlitz dutifully objected that only the doctor could decide on this. But Herr Klees offered to take me to the department doctor himself. He would dispel any concerns. We immediately went to the doctor with whom Herr Klees talked for a short time, and after just as brief an examination he had achieved his goal. We were both very happy with the result.

Of course, in order to be able to take up service in the merchant navy, I first had to be discharged. But that occurred with an almost magical swiftness. My discharge papers were issued that same morning. As when assigned to U 21, I was again envied by all comrades; nevertheless they wished me much heartfelt congratulations. Leave taking remained to be done and everything else had to happen chop-chop. I was given a free ticket with the discharge papers so I could go home to get civilian clothes. I was in a great hurry because U-Deutschland was supposed to leave for America in a few days. I myself was so concerned I immediately took the afternoon train home. My parents were astonished when I suddenly stood in front of them without prior notification. I quickly explained the matter to them. They were also happy I could now join the voyage on U-Deutschland. During the remaining hours I unfortunately had ample opportunity to convince myself the mood here in my father's house was completely different from that of my first leave from Pola. Food had become scarce. My parents were very upset by the circumstances and very old. Besides me, four brothers were drafted into military service, and the worries about them weighed heavily on everyone. I promised to bring back many beautiful things for them from America from the trip.

When I returned to Kiel, I immediately reported to the new duty. For the time being, it consisted in having to familiarize myself in detail with the technical arrangements of U-Deutschland. In doing so over the last few days, I had the privilege of receiving some amount of the care packages one received for crewing U-Deutschland. During this duty I realized the tremendous progress submarine construction technology had made. When I thought back to the little boat on which I had learned, or of the first trips of U 21, it seemed almost incomprehensible that in the space of a few years such tremendous progress was made as the result of which the miracle of U-Deutschland went to sea. Previously the mechanism of the small submarine seemed comparable to complicated clockwork, and now compared to this miracle construction clockwork was a downright simple thing. It would be simply amazing, as I immersed myself with the greatest concern and interest in submarine construction, to be able to familiarize myself with my new field of work in the short time that was still available. Because there were only a few days left, as the second trip was due to start at the beginning of October. I was able to get used to it quickly thanks to versatility in the service before August 1914 and from August 1914 to April 1915 in the shipyard and the later duty on U 21.

On October 2, 1916, U-Deutschland set out on the journey. We went through the Kaiser Wilhelm Canal to Wilhelmshaven. Here we had to go to the shipyard because of damage to the hydroplane and stay there until October 8th.[38] On this day the long journey finally started. In doing so, of course, a path was taken that expected the fewest dangers and encounters. Because U-Deutschland was not going to any fight, but only had to bring its precious cargo safely to its destination. Therefore we drove as close as possible to the coast of Schleswig-Holstein and Denmark, past Copenhagen and a good bit to the north on the Norwegian coast. Only then did the dangerous part of the journey begin - across the northern part of the North Sea. Because here we had to go through huge minefields, so that the utmost attention was required to reach the open Atlantic in spite of all the dangers. There was nothing dangerous about the journey across the ocean. We avoided every sighted steamer in good time so we could not be seen or hunted. The lookouts held the most important duty. We always depended on their attention so we could go into hiding before an enemy eye spotted us. Any armed vessel would have shot at us without further ado. The outbursts of fury in the enemy press and the threats made by enemy statesmen after the success of the first voyage were vividly remembered. We knew too well we would be hopelessly lost if an enemy succeeded in confronting us. Therefore, not only the lookouts paid the sharpest attention, but also all those on watch in order to obey every sudden command as quickly as possible.

Other than that, life aboard this merchant submarine was extremely pleasant in every respect. Besides me, there were five assistant engineers on board. The engineering watch was divided up so after four hours of engine room duty each of us had eight full hours of rest, for which we had a nice living quarters with equally nice bunks. An excellent cook knew how to prepare tasty dishes. The food was

[38] A harbor tugboat collided with U-Deutschland as the latter was swinging away from the dock damaging the starboard diving plane.

therefore excellent. It almost felt like I was on a vacation trip; because on my previous submarine trips I was used to something different. I told my comrades a few things about what happened on a war boat. They could hardly believe my descriptions since none of them knew this duty from personal experience.

We mostly made the trip to America surfaced, apart from the occasional dive and the subsequent underwater travel. However, we were held up many times by heavy seas and we needed twenty-four days for the crossing. When we finally saw land, we first did a trial dive for a few more hours, and afterward were then to approach the land so many longed for. I was of course full of expectations. Most of all, I was looking forward to surprising a friend of mine named Kurt Feine with my arrival. Kurt Feine was part of the crew of U-Bremen. And U-Bremen should now be lying at the Dollar waterfront for eight days already.[39] I met my friend Feine one day in Kiel before I even intended to serve on U-Deutschland and he told me he was part of the crew of U-Bremen. He recommended I report to this ship too, since they happened to need an experienced submariner. However, I refused this suggestion and at the same time asked him not to try any further, since I had no intention of doing another service on a submarine. He regretted this very much, but saw my mind was made up and did not press me any further. It was actually my firm intention not to be active on a submarine any more. And now I had gone along after all.

We came increasingly closer to land. It came to life in the boat. Everyone was looking forward to getting solid ground under their feet again. After midnight on November 1, 1916, we entered the port of New London. We stopped in the middle of the harbor, the siren roared several times into the night to signal our arrival. The navigation lights were lit and after a short time we heard a motorboat departing from land and approaching us. We were called from afar, asking if we were the "Bremen". Captain König called back from the command tower: "No, we are U-Deutschland." In the meantime the motorboat had come very close to us. The previous

[39] U-Bremen had departed Germany for the USA in September 1916.

voice rang out again: "Ah, Captain König, then you are the ..." "Well, then the 'Bremen' probably no longer exist. Two steamers have been searching continuously for eight days and cannot find her." We were shocked at this news. That was really not the reception we had imagined it to be. My first thought was of my poor friend. I saw him standing in front of me again, as he had told me with great joy in Kiel that he was going to America with U-Bremen. And now he would no longer achieve his goal.

The gentlemen from the motorboat came on board to us. There was a doctor, a customs officer and the well-known Captain Hinsch. Mr. Hinsch was the organizer for the merchant submarines service in America.[40] He told us a reception party had been planned for the arrival of the Bremen and also for us, but this would now be canceled because "Bremen's" loss was to be expected. Further exploratory trips to find her are now pointless and will also be discontinued. After a short medical examination, we were directed to the new berth, which we immediately took. The boat was enclosed all around at this berth. Two large new storage sheds stood at an angle to each other. On the third side was the steamer "Willehad". Between the shed and the steamer was a floating fence which could be opened and closed.

The cook from the steamer "Willehad" made us a strong coffee and delighted us with nice fresh pastries. We first strengthened ourselves thoroughly and then went to rest for a few hours. The next morning work started again early. The crew was tasked with thoroughly examining all aspects of the boat. The diesel engines in particular were carefully and meticulously overhauled, because the success or failure of our trips depended on them. If they failed, then our fate was sealed, then we would disappear just as in likelihood "Bremen" had fared. Therefore, get the engines in order! That was the beginning and the end of our thinking. Of course, all repairs were carried out by the crew themselves. It was not for nothing that when they were put together, meticulous care had been taken to

[40] Friedrich Hinsch also operated as an agent provocateur spearheading sabotage activities within the USA.

ensure that only those personnel who could meet the requirements were selected. During this repair work one could see how important training and instruction had been in peacetime. Without such thorough training, it would not have been possible to assemble together many crews.

While we were checking the boat's technical equipment and, if necessary, fixing it, a thick column of negroes was busy loading and unloading cargo. This work was directed by a very skilled specialist. Because it was not indifferent in which way the cargo was stowed. Care had to be taken to ensure the boat was always in the correct trim on the return journey and especially when diving. Whole mountains of semi and raw rubber disappeared inside the boat. Even to us, who knew our boat, it was a mystery where and how these huge quantities were housed. The foreman of these workers loading was must have been in league with the devil. We were told the entire German army could be supplied with rubber for cars and motor vehicles for four months with the material we now took on board. From this one may of sorts infer the enormous importance of these merchant submarines.

Our work ended at 5 p.m. every day. Afterwards we could go ashore every day and spent the hours in new festivities. I did not expect the Americans to be as enthusiastic as we are of being here. The cars who were supposed to pick us up after work were already queuing in long rows at the port. We were literally fought over. Everyone wanted to take one of us as a guest. We had free access to all clubs and theaters. German families invited us always as welcome guests. I often and gladly accepted such invitations. Again and again we were asked about how things were going in Germany. Everyone was happy to learn something from the mouths of Germans and fellow countrymen about the conditions in their former homeland. Here in America one had to rely on English newspaper accounts. One was convinced these were not always correct, but had no real opportunity to get a picture of actual conditions. They were extremely happy to finally hear the truth. It was really good for us to see how these people grew fond of their homeland here in a foreign country, how they worried and how they feared for the fate of Germany. And they transferred all their boundless love of home to us. They really couldn't offer us enough. One day the crew received an invitation from a German club in a neighboring town. The chairman picked us up. We only had to travel a short distance by train. When we arrived at our destination and climbed out of the train, we were presented with a grand sight. A large band stood in front of the train station and played Deutschland, Deutschland Über Alles, countless people burst into enthusiastic cheers when we appeared, young women thrust flowers at us, and tremendous cheers roared around us. We were completely surprised by this unparalleled warmth. We hadn't seen anything like it before. We first formed up behind the band and a huge crowd joined in. This unique festival marched to the clubhouse, where a large table was festively set in our honor. During supper, speeches were exchanged back and forth; lectures of all kinds followed, the band and choir vied with musical performances. In short, there was a real German festive mood. If you hadn't known you were in America, you might have thought you were sitting somewhere in your home country.

On my last brief visit at home, my mother had given me the address of an uncle I didn't know, named Eichelmann, who was supposedly living in Philadelphia. Immediately after our arrival I wrote to this uncle that I was lying with U-Deutschland in the port of New London and asked him to visit me one day. I also received an answer, but he did not want to come and asked me to go to Philadelphia. But there was a problem. I asked Mr. Klees to give me a few days' leave. But he refused, remarking that only Captain König himself could authorize. He, Herr Klees, doubts I would get leave now. I'd better wait until the next trip, and then he would be able to get me the three days vacation I wanted. But since I eagerly wanted to go to Philadelphia right now, I discussed with my comrades how the thing could be done without the knowledge of our superiors. After careful consideration, they agreed I should go. Somehow my absence could be covered up. Of course since I could rely on my comrades, I did not think twice about staying away for three days.

On a Friday evening at 8 o'clock I took the train to New York. I arrived at midnight. Here I sent a telegram to my uncle that he should pick me up at the train station. When I left the station concourse, a German, who might have recognized me as a compatriot by my clothes, called to me in German asking where I was going. I replied I had to go to the Pennsylvania train station. Since he was a cab driver, he immediately offered to take me there. I was quite happy about the fortunate coincidence. I was relieved of two dollars upon arrival at my destination.[41] A driver who made the same journey with another passenger and who recognized me as a compatriot by my uniform asked me what I had paid the driver. I told him the price. What? You really fell for it he said. I would have brought you here for eighty cents.[42] So the dear, friendly compatriot had been a kind of swindler.[43] I arrived in Philadelphia at 4 a.m. after several hours of traveling. But there was nothing to be seen of my uncle. I stood in front of the train station for over half an hour, completely disappointed. But when no one showed up to solemnly greet me; I got on the nearest electric tram to somehow reach my destination. After a few trips back and forth, I was dropped off by a conductor, and direction given to me to understand that I still had to go travel a few more streets to get to Hilton Street. So I went on a search. I had already walked a few streets without reaching the desired destination. Then I saw two gentlemen to whom I turned to question. I had hardly spoken when the older of the two clasped my neck with shouts of joy. It was my uncle whom I had unexpectedly met. My hands still hurt me today, when I think of how he shook them again and again beaming with joy, in a truly American manner. The younger one was his son, who also greeted me most happily. On the way to the apartment the two told me they had been waiting for me all night at the train station closest to them. But I got off at the first Philadelphia train station. My relatives, who were familiar with the area, had not thought of this possibility, while I,

[41] About $49.90 in 2020 U.S. dollars.

[42] $1.91 in 2020 U.S. dollars.

[43] Schultz uses the picturesque Bauernfänger, whose entomology traces to "Peasant catcher"

for my part, had neglected to inquire whether there were several train stations and at which I should get off.

In the meantime we reached the apartment. Again, the most joyful greeting. But since everyone had waited all night for my arrival, they were tired, and it was agreed to take a rest. But the rest didn't last long, and soon we were sitting around the breakfast table again. And now it was really time to tell the story. All possible and impossible questions were asked. The interest in conditions in the old homeland was quite considerable. But it did not stop with this celebration in the small family circle. The Eichelmanns had invited their entire circle of acquaintances for the afternoon. And everyone came because they all wanted to see and speak to the man from U-Deutschland. The next day, on Sunday, my uncle took me to his local pub to introduce me to his friends. The German-Americans present couldn't do enough to make me happy, but Americans themselves, even the French and English, joined in to honor and celebrate the man from U-Deutschland. Everyone wanted a little souvenir of me as well. But what? Then it occurred to me that I had a number of iron ten pfennig pieces with me that I gave as a souvenir. But they didn't want them for free. I was formally forced to accept an equivalent in money. I was unable to defend myself against it. When the ten pfennig pieces were over, German cigarettes had to serve; they too aroused a collecting zeal, and for each piece, as well as for every iron ten pfennig piece, I received a shiny dollar, so that I deeply regretted not having much more with me.

What impression the visit by a man from U-Deutschland made is best shown by the article that appeared about my visit in the major Philadelphian German newspaper and which I am including here for the sake of interest. It read:

> "On Saturday, the 11th of the month, an old, genuine German from our city, Mr. Josef Eichelmann, from 702 Hillion Street, was surprised with a special honor when a nephew of his, who had just come from Germany, visited him. A young, stocky German too. How did he come over

now? You will ask dear reader. Yes, that is now very simple, namely on the pride of us German-Americans, Deutschland, in the capacity of engine attendant. Well, it was a handshake and congratulations from all sides, even real Englishmen, who couldn't deny our brave, now world-famous blue jackets their respect. And then Mr. Josef himself, always a proud smile on the train[44], which only disappeared when duty called back the hero. Ms. Eichelmann, like a good German housewife, had of course made every effort to ensure the party was appropriate and there was no shortage of delicious food, beer and wine. Unfortunately, the fun didn't last long, because at 6 a.m. on Monday morning, Mr. Paul Schultz, the young hero, had to be back on board his brave little boat. What does Mr. Schultz says about Germany? He says Germany is currently still eating despite the starvation politics[45], even if it has to ration itself very much, and everyone is convinced of German victory. When asked whether the crew was not afraid of being caught, he assured them with a contemptuous smile that with their "König" they slipped through the clutches of the devil, how much more difficult can be those of the English thieves.

But we, who do not need to do such heroic deeds, or unfortunately cannot do them, wish this hero a happy journey and good-bye and a strong German Seaman's wellbeing."

On the way back I occupied myself for a long time with thoughts of those happy days. But the closer I got to my goal, the more I thought about the boat and the leave I had taken on my own. Had my two-day absence been noticed? I have to admit I was very worried after all. On Monday morning the train arrived in New

[44] Most likely a reference to Mr. Eichelmann's profession, possibly a train/tram conductor.

[45] Reference to the Entente blockade of any foodstuffs being imported into Germany; starvation being an economic and political weapon.

London on time. I quickly got on board. My comrades received me with joy and assured me everything was clear and in order. I thanked them warmly for their help and the camaraderie they had once again shown. Even if my conscience was a little squeezed at the moment, I did not regret later that I went against regulations at my own risk. Otherwise I would never have come to Philadelphia, because this trip was not only my first but also my last on U-Deutschland.

The loading work was gradually coming to an end and the crew was now making ample purchases. Everyone wanted to make their loved ones happy. Everything was available here in America, and on my last visit at home I had seen what was mostly missing. So it was no wonder that my purchases filled a sizable box.

Before we left, we learned from fishermen they had seen U-Bremen about eighty to one hundred nautical miles from land. When they arrived at harbor with their fishing boats and asked about the merchant submarine, they were astonished to learn U-Bremen had not even entered. Thereupon Captain Hinsch had two tugboats search for U-Bremen again. But all effort was in vain. U-Bremen remained missing. We only had one explanation for this: The boat must have made a trial dive before entering, and some disaster happened in the process. It is possible the depth control failed, and it is also possible a bulkhead broke through in the hold while diving, causing the cargo to fall forward, so that the boat's dive could no longer be controlled. Somewhere, not far from the American coast, it rests with its crew on the seabed.[46]

On November 17, 1916, U-Deutschland started the return journey. The main aim was to escape the lurking enemies who lay in wait on the border of American territorial waters with a large number of warships, submarine traps, nets and other diabolical tools of destruction. The battle to destroy U-Deutschland was to be waged with all means. If this time the ship came through the enemy blockade again, England's sea power would be exposed to the

[46] The fate of U-Bremen is still a mystery.

mockery of the whole world. At the initial departure of U-Deutschland, experts at the British Admiralty's staff had assured the world that a commercial submarine would "never" reach America. England's fleet would take care of that. And when, in spite of all these confident assertions, U-Deutschland made its way to the United States for the first time, it was trumpeted all over the world with even greater certainty that it would inevitably be destroyed when it left American ports. But nothing came of its destruction as U-Deutschland got safely back home.

Albion was never at a loss for a lame excuse. The "pirates" had just used some insidious ruse by which one had been deceived. Once and never again! The trip will not be successful a second time. But despite all the big words, U-Deutschland had already come to the New World. Now it was time to gain the way back. Again and again we were asked whether we believed we could escape the enemy again, and we were shown – often with grave concern – the newspaper articles stating with astonishing certainty of our impending doom. But to all questions and concerns we could only answer with happy laughs. None of us thought for a second the enemy's vile plan of combat against a merchant boat would succeed. We could rely on our captain.

So came the memorable day of departure. No doubt more dangerous than the day of departure from the home port. Because at that time the departure time could be kept secret, but now the departure time was known exactly to the minute. At the farewell, the enthusiasm of the Americans blazed up again in bright flames. Countless people came to wave to the crew of U-Deutschland and to see the ship again one last time, which seemed to be doomed for so many.

Before the departure, a test dive was made, which met our complete satisfaction. At midnight, the enclosure opened, and the boat left its berth. On leaving, the crew of the steamer" Willehad", which had served us as a barracks ship during our stay, called to us the warmest farewell and home greetings. Some of them would have liked to go with us. But that was not possible. Two American tugs gave us escort out of the port and accompanied us quite a distance through territorial waters. On one of the tugs was Captain Hinsch, who insisted on staying in close contact with us as long as possible.

We were already quite a distance from the port of New London when Captain König believed he could now continue the voyage without the escort ships. So he gave the two steamers [47]the signal to return. One of the steamers immediately turned and returned back. The second steamer, however - instead of turning outwards – moved in front of us, and before it could be prevented, U-Deutschland rammed the steamer and cut it into two parts with such rapidity that within a few seconds the whole ship disappeared in the depths. This accident happened so incredibly quickly none of the crew could be saved.[48] Only Captain Hinsch, who was alone in the foredeck, jumped overboard in time. By shouting he made himself known in the dark, whereupon life rings were thrown to him, with which he was pulled out of the water. During the accident I was on duty in the electric motor switch room, as we were still using the electric machines. Suddenly I felt a short jolt and at the same moment the command rang out: Both motors emergency power astern. I executed the order and then burst on deck to see what was going on. Captain Hinsch has just been fished out of the water. Of course there was great commotion onboard. There was already nothing to be seen of the rammed steamer, although the water was assiduously searched with searchlights.

What now? First our boat was searched to see if it had suffered in the collision. Unfortunately that was now the case. The ship had gotten a hole in the forward tank, which made diving impossible. Captain König therefore decided to return to the port with U-Deutschland. After the collision was properly reported, U-Deutschland was immediately seized and sealed by the American authorities. Captain König had to pay a large amount of bail until the court decided who was to blame for the accident.

---

[47] Schultz indicates steamers although these are tugboats as referenced in the previous paragraph.

[48] Five fatalities on tugboat *T. A. Scott, Jr.* built 1882 and only 62.5 feet in length.

The case was dealt with quickly by a competent court. During the inquiry it was perfectly clear U-Deutschland was in no way to blame for the collision. The boat was immediately released.

So there seemed to be a nasty ominous star hovering over our departure. There was secret rejoicing among the enemy and genuine concern among our American friends. But for us there was no wavering. We have to get through and we will get through, that was the thought animating us all. Our confidence in a successful return was not shaken in the least. First it was a matter of repairing the damage suffered. The necessary work was all carried out by the crew themselves. We were able to finally start our return journey on November 21st. This time the departure went smoothly and the numerous enemy ships assembled on the border of American territorial waters did not succeed in putting the slightest obstacle in our way. During the rest of the journey we had consistently favorable winds, and four days less was needed for the crossing. On December 10, 1916, we arrived in Bremerhaven, greeted by a huge crowd with boisterous enthusiasm.

Soon after docking, the cargo began to be unloaded. The crew, for their part, took care of checking and repairing the technical equipment in the usual way. As always, the greatest care was taken with the diesel engines.

The holiday season was gradually approaching, and with it the leave question became acute. It was out of the question for the entire crew to be given leave at the same time. The boat leadership divided leave takers into two groups; one could go home for Christmas, the other after their return. I myself belonged to the second group, which I deeply regretted because I would have loved to have been there when the gift box from America was opened. It contained twenty pounds of fine vegetable fat, ten pounds of butter, five pounds of coffee beans, five pounds of sugar, fifty bars of soap, two cans of olive oil, and lots of other bits and pieces. In order for the box to arrive on the festive days, I sent it from Bremerhaven as express goods and stated sewing machine parts as the contents on the delivery note. I didn't dare to give the correct content. With the bitter shortage of food, I didn't believe the box with this description would have arrived. Hunger hurts, and the whole of Germany was already starving at this time. But despite the timely dispatch, the box did not arrive for the Christmas festivities, every few days my relatives wrote me worried letters. So the festivities were over, comrades from the first holiday group came back, and the second group could now go home. But the reunion was much less joyful than before. When I first asked about the box, I was sadly told it still hadn't arrived and how essential the contents would have been. Both of my parents and the siblings were emaciated and starved, their previous rosy faces pale and sunken beyond recognition. Under these circumstances, of course, leave turned into a torture. I would have preferred to immediately return, so as not to take something from the little that was available while staying here. Only the urgent requests of my parents could persuade me to stay the three weeks' vacation time as an annoying deadweight. With a heavy heart I drove back to Bremerhaven afterwards. Here I immediately went to the place where the box had been left and requested an investigation be made into its whereabouts. I had little hope, however. But after a few days I received a letter from my relatives, in which they informed me the box had finally arrived. When the things were all unpacked and placed on the table, they were so pleased and happy that my old parents and my sister performed an Indian dance around the table. After all the grief of the past few weeks, I was pleasantly surprised by this news. It would probably be enough until the next trip out, and then an even larger food box would be brought back.

V. From U-Deutschland to U155

U-Deutschland was ready for the third crossing to America in the first half of February 1917. On the evening before departure, we were no longer given any shore leave and the entire crew had to stay on board. The final preparations for departure had been made and we were just waiting for the appropriate orders to cast off. Captain König then received a telegram with the instruction: "Do not sail, wait for an update." We found out more about it the next morning. It was America had declared war.[49] We were all extremely surprised because we had not and could not expect anything like this. After our experiences in America we all would have thought anything else was more likely other than a declaration of war. And how did we prepare for the planned third trip! Above all else, we had acquired a large supply of iron ten-pfennig pieces to exchange for dollars. They were crazy about it over there. We also had plenty of other items on board as souvenirs of U-Deutschland. All for the laudable intention of earning a decent chunk on this trip to be able to buy groceries. Our hopes and plans were naturally shattered when the surprising announcement of war's declaration was made. Too bad. But what now? What is happening to us now? What about U-Deutschland?

We didn't need to be in doubt for long. The cargo intended for America was immediately unloaded. On February 15, 1917, the remaining crew was paid off by the shipping company and we transferred into naval service. After unloading, U-Deutschland came to the shipyard in Wilhelmshaven, where it was converted into a warship. The crew was stationed in Kiel during the renovation work period. The reconstruction of U-Deutschland was

[49] The USA did not declare war on Germany until April 2, 1917. Schultz is probably referring to the February 3rd announcement whereby the USA severed diplomatic relations with Germany.

finished on March 21st and on that day we were ordered back to Wilhelmshaven. From now on, U-Germany was called U 155.

Very significant changes had been made inside the boat in the short time. The previous cargo holds were now used as living quarters and, above all, to store ammunition. Four torpedo tubes and two 15 cm guns had been installed.[50] As a result of this ample armament, it was no longer called a submarine, but a U-cruiser. The merchant submarines still under construction were also converted into U-cruisers, as their original purpose after the American declaration of war was no longer possible. These new U-cruisers were capable of great performance. They could make trips three times as long as the smaller submarines. Afterwards we did cruises of almost four months.

Since U 155 was now a warship, a purely military command atmosphere naturally prevailed on board. Some comrades who had not yet served in the Navy found it very difficult to get used to this. Least of all those who had served as engineering assistants on the merchant boat Deutschland and were now only stokers or senior stokers. But it was war and not everyone could be asked what was right for him or not.

But I would not be allowed to take part in U 155's first war cruise. During all the preparatory work, I was tormented by a persistent sore throat that worsened from week to week. Eventually I almost completely lost my voice and had to go to the hospital, where I stayed for over two months. Even after my release I was not yet sufficiently recovered to be able to resume work immediately, but had to stay in the barracks for several weeks. In the meantime I had been promoted to machinist's mate.[51] Since I have long earned this promotion, I always expected it. Still, I was very happy about it. However, it was clouded by the fact that I was no longer assigned to U 155, but to another former merchant boat, U 151. It too had for some time been converted into a cruiser but could not yet go to sea

---

[50] Provided by the pre-dreadnought battleship SMS Zähringen.

[51] Maschinistenmaat

because only a small part of the crew consisted of trained submariners, and a long period was necessary to train the new crews. So the whole summer and the beginning of autumn went by until U 151 was finally cleared for sea, but now things turned out differently than I had thought. Shortly before the departure date of U 151, my dear U-Deutschland returned from a long war voyage as U-cruiser U 155, and I was able to greet my old comrades again. It was immediately clear to me and my comrades that I absolutely had to come back to U 155. But that wasn't easy, because naturally they didn't want to release a man from U 151 who had been working on it all these months. Only with all sorts of trickery and cunning did I manage to get away from U 151 just before its departure.

And now shipyard work began again on U 155, which lasted until the beginning of 1918. It was not until mid-January 1918 that U 155 started a another long-distance cruise again. On January 14th, at 9 a.m., everything was ready for departure. The fleet commander appeared in person and gave a short speech to the crew: This time the boat was destined for a special important mission, for which he wished us good luck and complete success. U 155 was now under a new commander. The former had been replaced by Lieutenant-Commander Eckermann[52]. Before the departure, the last diving attempts were made in the bay of Eckernförde, and then the journey continued to Travemünde. We arrived there on January 15th. During the day, practice drives were carried out again, and in the evening the journey continued.

In the entrance to the sound, S.M.S. "Hannover" lay as a guard ship, surrounded by huge minefields. All submarines leaving for a long-distance voyage first had to dock at the Hannover. From here they were led by a torpedo boat through the minefields extending into

[52] Korvettenkapitän Erich Eckelmann, (1880-1929). Recently graduated from the submarine commander's course without having previous service in submarines. U 155 was his first combat command.

the Sound.[53] We also moored at the watch ship and took this opportunity to take on drinking water again.

It was known English submarines lay in wait in the Sound. Extreme care and vigilance was therefore required. U 155 was towed by a torpedo boat so it would not unexpectedly come across a mine on its passage through the Sound. A second torpedo boat zigzagged ahead of us to prevent attacks by enemy submarines. Despite the great danger of mines, the journey went very quickly, as both the torpedo boats and we drove with extreme engine power. The torpedo boats turned around before Copenhagen. At 8 o'clock in the evening we glided so closely past this beautiful city we could have reached it with a stone's throw. After only two days we were in the North Sea. Here again extreme caution was required.

The entire sea was contaminated by the English with minefields. In order not to get into one of the fields or contact one of the numerous English patrol vessels, we stayed close to the Norwegian coast - all the way beyond Bergen.[54] Because of the dangerousness of the area passed and the unnecessarily heavy swell, but also in order to be able to dive at any moment, no crew member was allowed on deck for five days, and we had to remain on "high alert". It was only after we were out of the danger zone that we were allowed on deck again. Enjoying the fresh air was a real relief. During the drive through the danger zone, the Watch was divided into two groups, which replaced each other, i.e. half of the crew was on duty while the other half was resting. Now that we were in the open sea, the watch duty was carried out by three groups, which resulted in a longer rest period for a third of the crew.

The caution with which we had to drive all the time meant we needed a full nine days to cover the route across the North Sea from the coast to reach the open sea. But even now the journey was by no means smooth, because a new obstacle was in the way. The sea was

---

[53] Fehmarn Sound is the waterway leading to the Fehrman Belt, the strait separating Germany from Denmark.

[54] Misidentification by German submarines was also a threat as U 155 was nearly torpedoed by U 19 on the previous voyage in these waters.

very choppy and the waves kept tossing us back and forth. We were forced to proceed slowly, the diesel engines were switched off and only the electric motors were used.

The storms stopped us for several days. It was not until noon on Saturday, January 26th, that the weather cleared up so that at least one of the diesel engines could be put back into operation. Even so, we made slow progress. In the evening, however, the sea had calmed down enough so both diesel engines could run again. Our commanding officer immediately took the opportunity to do a test dive to see whether everything was okay despite the stormy weather. There were no defects. The next day, the usual service was given a pleasant change with the Emperor's birthday celebration. In front of the assembled crew, the commandant gave a speech adapted to the celebration, which ended in three cheers for the Emperor. As a special festive allowance, there was a delicious cake baked by the ship's cook in addition to the other meals.

Singing did not for long maintain the good mood of the weather gods. Over the next few days the sea was always so stormy we only made very slow progress. Eventually the swell became so strong the commander decided to proceed underwater. First an attempt was made to gain better speed at a depth of thirty meters. But even here we were still too affected by the waves. Even at thirty-five meters deep we didn't feel much relief. Only at forty meters was it possible to properly steer the boat. So we stayed at this depth until evening. Then we surfaced again, but even now we hardly got anywhere. After eight days of driving we were only at the latitude of Scotland. It just would not go forward. The storm and the countercurrent of water continued for several days. At last we got the sea on our side, and as a result the voyage immediately quickened. We were able to put both diesel engines back into operation. Our area of operations required a southern course. It wasn't until February 7th, when we were near Cape Finisterre, that the weather improved significantly, which finally brought some relief for us. Because the whole time U155 struggled with the unusually strong waves and no one was allowed to come on deck. The stay in the boat had therefore become almost unbearable. The little fresh air that could get into the interior of the boat through the tower hatch did not bring the slightest relief from the smells and vapors produced by oil combustion. Because the boat was almost completely closed, there was no ventilation. The various vapors and constant handling of oil caused everything in the boat to become damp and slippery. The heat was unbearable. In the engine room, men worked only in very light trousers. Those who could were on deck to pump their lungs full of fresh air and to get rid of the smell of oil. Of course, the boat itself was also thoroughly ventilated, making the stay in the engine room more bearable again. With the favorable weather we made rapid progress. And now we also found out what our task was: we were to cut the submarine cables near Lisbon. An operation that made the highest demands on commander and crew. The cutting of the cables was only possible immediately in front of the port exit, so the risk of being seen and of sudden gunfire was extremely high. Only a leader with lightning quick decision making could accomplish this task, and only with a crew whose training and attentiveness knew how to execute every order just as quickly.

We proceeded slowly and carefully when we came near Lisbon on Sunday, February 10, 1918. Until then we had apparently gone completely unnoticed. We evaded a steamer we spied in the distance by diving quickly. At 8 o'clock in the evening we were at a point off the port of Lisbon where the cables were. U155 looked for a convenient place between two unused lighthouses. Our commander thought the moment to act was opportune. After all preparations were completed, the cable shears were lowered and the two diesel engines started. With the ship moving quickly, the cable shears were dragged on the seabed, and the cables would be grasped and cut through. The cable cutting machine was connected to a measuring instrument, from which one could read when the cutter encountered resistance, i.e. when a cable was detected. According to our observations, three cables were cut that night. Then suddenly the cutters no longer took. Something was wrong. As the wire rope was being wound up, we noticed, to our unpleasant surprise, the cutter had torn off. So we headed out into the open sea again to attach a new cable cutter undisturbed. It took so much time we couldn't finish our work that night. So we stayed in front of Lisbon all Monday, naturally in a manner so we couldn't be spotted.

Only after dark did we return to the scene of our nocturnal activity. The remaining cables should also be cut. When we arrived at our destination, we immediately started work. The cutter was lowered again. Immediately after the diesel engines cranked, there was a jolt and the meter indicated a high load. But the engines pulled through strongly. After a short time we again noticed that the second cable cutter had also torn off. And that put an end to our activity. We left the harbor and went back out into the sea. The next day we were still outside Lisbon. A Spanish steamer was stopped and its cargo examined.[55] He was able to continue on his way since he had only non-contraband goods on board,. On Wednesday our commander assembled everyone up on deck and announced the success of the cable cutting: four cables had been cut and a fifth badly damaged. We were satisfied with this result. We found out later from our enemies that these five cables were correct when we intercepted a radio message confirming the cables from Lisbon had been cut.[56]

Following the announcement of our work's accomplishments, the captain told us what the next destination of our voyage was: the Strait of Gibraltar and the area from Gibraltar to Madeira. In this are we could attack enemy merchant ships.[57]

On my previous trips on U 21, the fight had been aimed exclusively at warships. Destroying them had been our job. Now the operations of a U-cruiser were extended to merchant ships if they had loaded war material or banned goods of any kind. Of course, that was a much more difficult thing than releasing a torpedo from a safe

---

[55] Steamer Arnabel Mendi.

[56] The cable lines to Gibraltar, Falmouth and Sao Miguel were cut. The two to Madeira were not.

[57] Germany declared all shipping would be attacked without warning in certain war zones. The waters encompassing Madeira, Azores, and shipping lanes towards Gibraltar and passing through the region were declared a war zone on November 21, 1917.

underwater position on a vessel. Merchant steamers had to be stopped at sea. That alone was a big risk. How easily a shot could be fired at us from a stopped steamer, which would send the boat and its crew into the depths. And almost all enemy ships were already equipped with guns to defend themselves against submarine attacks. If we managed to stop a steamer and examine its cargo, there were several options. If the cargo consisted partly of goods for the enemy, this part was thrown overboard. If the entire cargo was for one of our enemies, or if it consisted of war material, the crew had to board their lifeboats and be exposed to waves on the open sea. Whether and when they would ever reach land could not be asked. It was war; we had to ensure our enemies had as little supply as possible. The steamers were then always quickly sunk.

We increasingly noticed we had arrived to a southern, warmer zone. There was gorgeous clear summer weather with a magnificent calm sea. Staying on deck was a wonderful pleasure. We spent most of our free time on deck and could relax in the lovely air. The downside was staying in the engine room became increasingly unbearable the further south we went. The temperatures in the engine room rose to forty-eight or fifty degrees.[58] Because the terrible heat in the boat, the watch was again divided into four groups: four hours of duty and eight hours of rest, provided no attack was made during the rest period. Then of course everyone was up and running. And that was the case often enough in the next few weeks. Unfortunately not always with success. For example we sighted a steamer on February 15. We immediately made preparations for a torpedo attack. It took a long time to setup the firing shot on the steamer. But the all the hours of maneuvering was in vain. The shot failed because the torpedo did not leave the torpedo tube due to an error. We then pursued the steamer through the night and were able to attack it with the guns at 5.30 the next

[58] 48°C – 118.4°F, 50°C – 122°F

morning. But after a long battle he escaped because he was able develop greater speed at full steam.[59]

Two days later we were luckier. An American sailing vessel was stopped on Saturday morning. He carried salt meant for the enemy. The crew had to leave the ship, and it was sunk with two shots.[60] At noon we again fought a steamer. Immediate quick dive. We attacked submerged with a torpedo shot and, to our delight, saw a hit. The steamer stopped immediately and launched lifeboats. Immediately afterwards there was a violent detonation because the engine room had been hit and the boilers exploded. In the meantime we surfaced again. The boats stopped for us when they saw us. We then learned three men had been killed in the boiler explosion and a man in the lifeboat was seriously scalded. The unfortunate man was bandaged by our ship's doctor. The sunken ship is the Italian steamer "Tea". It was 5,355 tons displacement and loaded wheat.[61]

On this lovely day our search for more prey was in vain for some time. On Saturday the Spanish steamer "Sardiniera", 2300 tons in size, was stopped. Also with a fine cargo of wheat destined for France. The crew therefore had to leave the ship. They grumbled and gestured constantly, but it didn't help. They had to get into their boats and we gave them a course where to row. The steamer was then sunk.[62]

---

[59] Steamer Craonne, 4,264grt. Pursuit discontinued due to heavy seas and lack of visibility. U 155 was still equipped with the lower horsepower 400hp Diesels fitted for merchant service, plus the additional weight of war armament and beamy merchant hull meant a relatively slow surface speed at 12/12.4 knots under the most favorable conditions compared to newer steamers. U 155 would prove to be outpaced by steamers rated at and less than 12 knots.

[60] The Nova Scotia registered schooner 102 ton Cecil L. Shave. Ostensibly British flagged and Schultz may be forgiven misidentification with American. There were no survivors among the crew of six. (Some source indicate crew of twelve.) She was sunk off the Azores en route from St. John's, Newfoundland to Gibraltar.

[61] The 5395 ton Tea carrying grain from Portland, Maine to Gibraltar. Sunk 80 miles SW of San Vincent. 1 fatality. Spindler, Vol 5, op. cit., indicates S.S. Tea was sunk on February 16 and the S.V. Cecil L. Shave on February 18. Schultz may have conflated the dates writing thirteen years later.

[62] The 2170 ton Sardiniero was enroute from Argentina to Sète, France with a cargo of grain. The ship carried a safe conduct pass issued in Berlin as the ultimate

On Sunday, February 24, 1918, at 8 o'clock in the morning, we stopped another Spanish steamer. When his cargo was checked, it was found that he had around 300 tons of contraband goods and other cargo for France on board.[63] The steamer's crew had to throw these enemy goods overboard under the supervision of our boarding party. The Spaniards reluctantly did this work and tried to prolong it as much as possible, probably in the expectation that help would come to their side, and possibly forcing us to disappear. Their stubbornness did not help them. There were also some Spanish officers on board the steamer, who invited our captain; but the latter flatly refused, remarking he wanted nothing to do with those who supported our enemies. Because it can be assumed with certainty the Spanish officers knew there was cargo for France on board the steamer. It was a strange scene with all the boxes and crates, barrels and bales floating around on the sea. We couldn't resist the temptation to fish one or the other object out of the water and bring it into the boat as we passed. We thought of the misery of our relatives at home and it was hoped this would bring them some relief.

At 9 o'clock in the evening the task was done and the cargo destined for the enemy was in the water. The steamer could now continue its journey. The next day the number of Spanish ships we stopped increased again. But this time the cargo was not contraband and the steamers could continue on their way without further ado.

The closer we got to the Gibraltar narrows, the sharper our already intense attention became. Not only was steamer traffic livelier than

---

destination of the cargo was Switzerland. The vessel was sunk due to alleged misuse of the Spanish flag some 20miles from Casablanca. The German government transferred an interned German steamer in Spain to the Spanish as indemnification.

[63] The Spanish steamer Villa Real was stopped about 30 miles from Casablanca and carrying footgear and foodstuffs to Morocco. The German prize officer inspected the documentation and informed the captain that one hundred tons of cargo he was transporting bound for the Moroccan port, consisting of food products and footwear was war contraband because he did not have the certification of the German consul. The officer demanded the cargo be thrown into the sea and for five hours, the crew and passengers who volunteered to do so, threw the cargo overboard. The captain had to promise that he would not make a stopover in Casablanca and was allowed to continue his journey to Santa Cruz de Tenerife.

on the vast ocean area – auxiliary cruisers from that English war port also ventured out a little more here.

Fortune seemed favorable to us. After several days of uninterrupted spying, we sighted an armed merchant steamer on the night of February 27-28. Immediate quick dive, torpedo readied for launch, and the enemy is quickly approached under water. At night one could be more worry-free in this context because the U-cruiser's periscope could not be noticed by the enemy. On the other hand, however, the prospect of scoring a direct hit was less. And this disadvantage also made itself felt here again. The torpedo shot failed and the steamer escaped without even having a clue that a submarine was on its heels.

A few hours later we ourselves were the quarry. It was not long before an enemy auxiliary cruiser spotted us and immediately began to hunt us. We disappeared by quick diving, assuming the enemy would soon turn away again if he had us out of sight. The course was changed underwater, so that we remained undetected during a cautious reappearance.

The next day there was another encounter with an auxiliary cruiser. This time our captain decided to attack although with the heavy swell there was little chance of success. After much back and forth a torpedo was finally fired, but as feared it went wrong and we departed without having achieved anything.

On March 3rd we were finally able to book another solid success. In the quiet of a late Sunday afternoon came the loud voice of the lookout on duty, who had seen a trail of smoke. Since an interception and a torpedo attack was not possible due to the distance and the speed of the enemy, the attack was carried out immediately with gunfire. Our fire was vigorously returned from the steamer, but without our suffering the slightest damage. The onset of dusk did not allow us to get a hit either, and the darkness that quickly followed in the southern latitudes enabled the enemy to escape heading for Gibraltar. He kept trying to get help with SOS calls. But he made the mistake of trumpeting his course and speed around the world with the calls for help, which our radioman caught with a happy smile. Because now we easily calculated we would overtake the steamer in a few hours if we drove with high speed. The pursuit was immediately taken up. We reached the steamer's intersection point exactly according to our calculations. Our boat dived and a torpedo attack immediately followed. The result was favorable; the steamer was hit, even if it did not immediately go down. But the crew left him and got into the lifeboats. Lying submerged, we watched the steamer through the night.[64] At dawn we appeared for a short while, only to quickly disappear again as we saw three destroyers approaching.

The new day brought us no rest despite the stressful night. Again it was an English auxiliary cruiser which, to our delight, set course for the Strait of Gibraltar. We immediately approach the enemy, let go a torpedo, and waited feverishly for the impact, which according to our calculations should be successful, since the sea was relatively calm. But this time the enemy was very alert, well warned by the attacks we had made in the previous days. At the last moment he turned off the torpedo's path and headed for us at full speed. Our captain refused to engage in another fight with the overpowering enemy, and U 155 disappeared in the depths.

A few days later we almost got into serious trouble due to machinery failure. We were in the process of chasing a steamer when suddenly a compressor failed, so we could only continue at a slow pace. Fortunately, our opponent was seized with such terror that he hardly cared about us, but sought his own salvation in the most hurried flight. It was a difficult job to correct the mishap and to get our boat back to normal performance.

Very soon we would see how good it was the failure had only happened to us while pursuing a rather harmless merchant steamer. Because on Sunday, March 10th in the evening, the Norwegian steamer "Wecadesk" was sighted and immediately shot

[64] The 4439 ton Antioco Accame enroute from Rosario to Italy. There were no casualties.

at because it did not obey our command to stop. Six rounds were fired at the 4270 ton steamer. Then at last it stopped and the crew got into the boats on which they tried to escape. But we drove after them and caught up with them in a short time. The master was identified and asked to clarify the cargo. From the reluctantly handed over papers it was established the cargo consisted of brass, which our enemies would use to manufacture cartridge cases.[65] Our captain ordered the enemy to return immediately to the steamer in order to bring as much of the cargo as possible on board. The enemy captain refused. He was given a short cooling-off period and threatened he would be shot if he did not comply. In response to such energetic language, the captain preferred to order his crew to reboard. The cargo, which is extremely valuable for Germany, was brought over to us boatload after boatload. We worked hard until late at night. Then a rest period was ordered and the enemy crew was warned not to do anything stupid during the night.

The transfer of brass continued on Monday morning. In the course of the morning we were disturbed in our work by the appearance of an auxiliary cruiser. Assuming some misfortune had occurred to the stationary steamer, the auxiliary cruiser headed for him. Then we immediately disappeared into the depths. It had to be done so quickly we had to leave behind the rather large prize crew[66] who oversaw the unloading work on the Norwegian steamer. In the depths we were very concerned about our comrades. We were convinced a German submarine crew would not fare well if they fell alive into the hands of the enemy. So we feared the worst. From the depths we observed the events taking place on the surface with great tension. At the same time, the captain had all preparations made to take up combat with the new enemy if necessary. He was determined to do his utmost to save his comrades. But it turned out differently. To our greatest surprise, we noticed the Norwegian steamer immediately turned on steam and continue its voyage

---

[65] The Norwegian flagged 4271 ton Wegadesk was enroute from Baltimore to Genoa loaded with 5000 tons of brass, roofing felt, and sheet metal. Captain Hans Hansen commanding.

[66] 14 men composed the prize crew.

without waiting for the approaching ship to arrive. When the auxiliary cruiser saw the steamer in motion again, it changed course and did not care about it. For our part, we followed the Norwegian underwater and were not a little astonished when he stopped his voyage as soon as the auxiliary cruiser was out of sight. We now approached him again and surfaced near him. The prize crew saw us and made themselves known, we then went close to the steamer. Now we also learned the solution to the riddle: as soon as the auxiliary cruiser came into view and we dived, revolvers were held in front of the steamer's crew with the measured order to continue the journey. Completely amazed and frightened, the crew complied without resistance, and in this way the Norwegian had quickly got off to a brisk start. On the auxiliary cruiser it was probably assumed the steamer had stopped due to some accident. The cruiser had been fooled due to the presence of mind and skill of the prize crew. Our comrades were happily congratulated by us on their success, and the brass transfer was resumed immediately. Now our entire crew took part; even officers helped with joy and enthusiasm. The transfer was terminated after we had forty-five tons of the precious metal on board. It was now Wednesday, and only now could the steamer's crew get into their boats and continue their voyage into the unknown. The steamer was sunk.[67]

On Friday, for a change, we caught a Spaniard again, the sailing vessel "Joagnina". He had cargo for the enemy and was therefore sunk after the crew was released.[68]

Sundays were apparently our lucky days because for a few weeks now, Sunday had always brought us big booty. It is no wonder that

[67] U 155's War Diary indicates 40tons of brass was transferred and then the steamer was scuttled with charges. The Norwegian crew reported about 50 tons, food supplies, navigational instruments, tools and a few other things were taken from the ship. The crew reach the Moroccan city of Spahi on March 15.

[68] The 333 ton schooner rigged polacca Joaquina was enroute from Cuba to Cadiz with mahogany and other precious wood. Captain Francisco Perelló commanding. Sunk for being 500 miles from Cape Espartel, thus within the Forbidden Zone, where all shipping was liable to be sunk. The ten man crew embarked in two boats and was rescued 26 hours later by a convoy enroute to Gibraltar. The Official German History incorrectly lists the Joaquina as being eventually salved. Understandable as the wood cargo kept the vessel afloat for a lengthy period.

on the new Sunday (March 17th) special attention was paid to see whether there was a smoke trail on the horizon. And sure enough it was. Again it was a neutral who came our way, the Norwegian steamer "Reidar," 3,570 tons, which of course immediately received the command to stop. He too did not believe in the seriousness of the situation and tried to escape. Fire was therefore opened on him immediately. Since it was already a bit dark, we couldn't notice much the effect of our shots. We saw the crew was leaving the ship. Out of caution, the steamer was not examined further because of the darkness. During the night we stayed close to him and did not leave him out of sight. At daybreak we drove around the steamer underwater, observing closely, but could not see anything unusual. We then surfaced and an armed prize crew was sent on board. Not a soul was on the ship. We were only left with one beautiful dog. The cargo consisted of flour, probably intended for our enemies, otherwise the crew would not have tried to escape.[69] We were able to easily supplement our own flour stocks from such abundance. While we were still busy, a second steamer approached us. Despite a fairly heavy fog, we saw two guns on board. Without the usual warning shots we immediately opened brisk gunfire, which was returned by the steamer, fortunately without hitting us. On the other hand, our shots must have had a strong effect. Because suddenly the white flag went up, whereupon our fire was stopped. Immediately afterwards the crew lowered boats, one of which was badly damaged by shrapnel, and rowed towards us. In the meantime the steamer began to burn brightly. Of the crew members who came over to us, one had a leg completely shot off, a second had shrapnel stuck in his lungs, a third had his jaw completely splintered, and a fourth had died as a result of serious injuries during the short crossing in the boat. The men were all taken on deck by us and the wounded were thoroughly bandaged by our ship's doctor. In the meantime the captain of the steamer said he had first thought our boat was a destroyer, so he was completely taken by surprise by our gunfire. The first shot from our side had

---

[69] The Reidar was enroute from New York to Italy with the flour. There were 4 fatalities.

been a direct hit. Five men of the crew were killed by it. Another shot wounded six men. The ship was the Italian oil tanker "Prometeo", 4374 tons.

The wounded groaned terribly. Only after our doctor bandaged them did they calm down. Our upper deck was full of pools of blood. It was a terrible and horrific picture: there the burning steamer and here the horribly mutilated in the red firelight. After our doctor had done his duty, we drove with the crew survivors to the abandoned steamer.[70] From here they were allowed to untie a second, intact boat, since they would not have gone far with their own badly damaged boat. In addition, they were supplied with provisions from our stocks, since they left their ship so quickly they had no opportunity to take anything with them. Then they were left to their fate and to try to get ashore somewhere with the seriously wounded. The Italian captain thanked us especially warmly for the active help we had given him and his crew. Soon the enemy was out of sight. Before we went any further, the steamer "Reidar" was sunk.[71]

We learned from the Italian we were at a so-called assembly point for steamers, from which ships would be safely escorted by a destroyer. So we kept a keen eye out to see if nothing would catch us unaware.

The upper deck, soiled by blood stains, was now thoroughly cleaned so we would no longer be reminded of our terribly suffering enemies. The ghastly scene had more or less got into everyone's minds. After cleaning we went to dinner, which we didn't really want to at this time. Our conversation revolved around the likely fate of the small boats. "Boys," said one "isn't that terrible about this

---

[70] The Italian survivors rowed to the Reidar.

[71] The 4017 ton Prometeo shipped 5500tons of naptha from Texas to Genoa consigned to the Italian Navy. The vessel was armed with two 76mm cannon. Captain Desiderio Tonietti (1881-1933) commanding. He was awarded the Italian Merchant Navy Silver Medal for Military Valor. Engineer Cadet Giuseppe Amato was posthumously awarded the Merchant Navy Bronze Medal for Military Valor. After eight days of rowing the two boats reach the Portuguese Cape Verde islands; one of the wounded died enroute. Total 2 fatalities.

submarine war? Now the poor devils are sailing the sea in their nutshells. No bandages and no help for the wounded, only a few days of drinking water. Under these circumstances what else may they have to endure when you consider their clothes and bandages are completely soaked by the salty sea? They will no longer see home if they are unlucky and not picked up by some other vessel encountered on the way." Our table talk was like that. "Aye, comrades," said another, "what you're talking about is very nice and worthy, but think about what happens when we get into such a situation. One good hit on our boat and we're all finished. But that wouldn't be the worst. Such a death would still be an honest sailor's death. But suppose one of us falls wounded into the hands of the other side. What do you think would happen to him then? Do you think they would treat us like we did to the Italians?" We had to agree with our comrade, it was just war, and it was simply you or me.

It was two hours after dinner before it was my turn to go on watch. I lay down on my bunk and tried to get some sleep. But I couldn't. Again and again I saw before me the distorted faces and mutilated limbs of the Italians. I was glad when I finally had to assume my watch. By working in the engine room one can focus on other things and have time to find one's way again.

As favorable as this Sunday had been for us, the following days were equally unfavorable. Nothing, absolutely nothing, was to be seen, no matter how intently every lookout tirelessly searched the horizon. Only on Friday, March 21st, did a large enemy steamer come towards us again. We opened rapid gunfire when he came within sufficient range. A real naval battle ensued. During the shooting, the steamer continuously radioed calls for help in all directions, but we persisted and pursued the fugitive, who was now taking a zigzag course. We fired over a hundred shots at him then we had to stop the chase because one of our diesel engines failed.[72] Soon

---

[72] The 6175 ton HMT Chupra was the vessel. Captain C.R. King (1874-1924) commanding. Chupra was hit once by a shell which made a large hole in the bridge, cut several steam lines and buried itself in sacks of wheat in hold #2 without

afterwards a cruiser and a destroyer appeared in response to the steamer's distress calls. We evaded their attacks by quick diving. We had better success the next day. At 6 o'clock in the evening we sighted the 3800 ton steamer "Avala". We dived and tried to attack him underwater with a torpedo. It was only after hours of maneuvering that we got the shot at 10.45 a.m. But as luck would have it, the shot failed. We then surfaced and chased the quickly escaping steamer all day and all night long. On Sunday morning we opened the fight with gunfire when it got light. Various hits could be observed apparently seriously damaged the steamer. Because the crew then abandoned ship just in time, as it sank a few minutes later into the depths. The watery grave had not yet closed over the steamer when we spotted a new victim, a sailing vessel who also anxiously tried to escape. After a short pursuit, we caught up with him. His name was "Gorgina" and he had loaded salt for America. The crew had to get into the boats and the vessel was finished with two shots.[73]

The following Monday also brought us good booty. We caught the Portuguese sailing vessel "Ria de Ave". He too attempted to flee when we unexpectedly surfaced, but our guns soon forced him to stop. His crew could only be induced by a prize crew to get into their boats, whereupon the sailing vessel was sunk by six explosive cartridges. Two days later the Canadian sailing vessel "Watauga" met his fate.[74] He was also sent down by explosive charges.

---

exploding. Capt. King received Lloyds Meritorious Service Medal for his handling of the ship during the 25 minute chase.

[73] The Italian 3751 ton Avala was enroute from New Orleans to Gibraltar. Captain Bartolomeo Maggiolo (1875-1918) commanding and posthumously awarded the Silver Medal for Valor.

The British flagged 103 ton schooner Jorgina operated out of New Foundland and was stopped360 miles NxW of Madeira. Captain Seph Osmond commanding. The crew of 6 rowed safely to the Madeira Islands.

[74] The 104 ton schooner Rio Alva was enroute from the Azores to Lisbon (Portuguese source) was stopped on March 25, 1918 100 miles from Madeira. Master Joaquim Manuel Machado commanding. The Portuguese reported the submarine as surfacing to leeward cutting off the chance to escape and partially dismasting the schooner with nine shots. The crew of eleven made shore at Porto Santo, Madeira Islands on March 31st.

In these eventful days, our "work" was made easier by favorable weather. For a change, Saturday brought heavy seas again. To everyone's delight, our commanding officer elected to have a rest day after the stressful days and weeks. Hatches were opened, engines shut down, and in complete calm the ship was left lifeless to the drift of the waves.

April 1st brought us a sad experience. The Portuguese sailing vessel "Lusitania" was stopped by us this day. His cargo consisted of cotton and steel goods. Therefore the crew had to leave the ship despite heavy seas. One of their boats was in very bad shape. But there was no other available, and we had to leave the crew to certain death, as they could not get far with this poor boat in this stormy sea. In view of the certain sinking, they clung to our boat with dear life. For our own safety, we were compelled to remove them by force, at which point they let out a terrible cry. As difficult as it was for us to use violence against the poor men, we had no choice. They too, were our enemies; they too provided enemy armies with war material which had to be destroyed. We were not allowed and could not be merciful. For a long time we saw the unfortunates floating on the waves in their tiny boats. We had no doubts about their inevitable fate in this sea. They too perished in the fulfillment of their duty.[75]

On Tuesday, April 2nd, our captain announced we would start our journey home in a few days. The diesel engines had to be overhauled in order to be fully prepared for the dangers that might arise. The sea area around the British Isles was still contaminated far out to sea with mines, minefields and submarine traps. English submarines were lying in wait everywhere to intercept German submarines. The utmost maneuverability of our boat was the first prerequisite for getting through the danger zones. Every control, every rudder had to respond to the slightest pressure. The diesel

---

The 127 ton schooner Watauga was on a voyage from Santa Pola, Spain and stopped450 miles WxN of Lisbon. 5 lives lost including Master.

[75] The 527 ton bark Lusitania (Lusitano in Portuguese) was enroute from New Orleans to Oporto with a cargo of cotton, iron hoops, etc. There were no survivors. The crew composed 27 men.

engines, as well as the electrical motors, had to work perfectly. The latter could only do so if they were properly fed by the diesel engines. On this account, special preparations were necessary for the journey home. In the open sea, the associated risks were low. Out here if something should happen to us, it would be coming into combat with a well armed vessel or we could be suddenly rammed. But that was well-nigh impossible, since sufficient care was always taken. It would be different on the way back, where we were threatened by hundreds of dangers we had overcome with difficulty on the way out and had certainly increased since then.

Overhauling the diesel engines at sea was an extremely difficult job. While one engine remained in operation, the other was completely taken apart. Every free space was occupied with dismantled machine parts. The work was made even more difficult by rough seas which threw the boat back and forth, up and down, causing the machine parts to repeatedly tumble over and over. But it had to be done. We called this work shipyard time spent at sea. During this time, of course, no prey could be hunted, as the boat was not fully maneuverable.

Another day of rest was ordered after the strenuous repair work. We glided quietly and peacefully through the waters. The off watch relax on deck with jokes and levity. One is dozing in sweet idleness, limbs comfortably stretched out, the other is thinking with a gloomy look of loved ones at home, yet another entertains his comrades with a mouth harmonica, to which some dance around. Of course, this picture "Sunday rest on board a submarine" does not lack the inevitable Skat brothers[76], nor the hobbyists who always have to do something, whose hands are not tired. If one observes this picture one could not believe we are at war, one could not believe the same men who enjoy their peace and quiet here do nothing more day after day, week after week, do nothing more than sail the seas, to sink one ship after another, one steamer after another into the depths. One could not believe the same men who celebrate their Sunday here so unconcernedly can possibly be sunk into the depths

[76] A popular three player card game.

in an hour or two or three in battle. But now it's quiet, now it's Sunday, a really good German Sunday. Everyone rests, everyone celebrates, everywhere are relaxed and simple, cozy people. How good it is to be able to relax your nerves a little after such unheard of tension. You don't have to always think about the alarm, you don't always have to be ready to dive or surface, you don't always have to operate the lever and wheel, you don't always have to be ready to fire a torpedo or shell.

In the midst of this Sunday calm, the tower lookout suddenly shouted: Vessel in sight. Now everyone came alive. Gone is the Sunday rest. Everyone was ready for action again in a flash. The captain immediately went to the command tower. He identified the vessel as the Italian auxiliary cruiser "Sterope". The boat quickly disappeared into the depths at the command quick dive. The commander watched the auxiliary cruiser through the periscope. "Well," he said, "he seems to be taking his time, but of course today is Sunday and he is in no hurry, as he glides comfortably through the waves, well, we'll perk him up. It's just a shame the sea is so calm. We can't unseen fire a torpedo at him underwater. But we have to approach, if not underwater, then on the surface. "And now a new command: "Surface! Gun crew prepare for action! Immediately to the guns after surfacing."

At the same moment we were already on surface. After one, two, three shots the steamer was startled from its contemplative calm. These were warning shots and a command to stop. But then it came to life over there. And what is that? The fellow had guns, he started shooting at us immediately and at a very brisk pace. So it was to be a fight, a serious fight of life and death. It was either him or us. At that moment a most genuine naval battle was in full swing, and gun discharges alternated in quick succession. The steamer offered our guns a bigger target than we did it. The effects of the shots were observed with zeal. To our dismay however, we found out we were not fired at from two but from four guns and a revolving cannon.[77] The more numerous enemy shells struck the water increasingly closer to the boat. We hadn't hit the steamer yet when shrapnel pierced our diving tank. Things were starting to get bitterly serious. Our men worked the guns doggedly. What? They want to sink us into a watery grave on the way home? Never! Look at that over there. Hurrah, a 15 cm shell struck the Italian's deck. Various parts were torn away. A second hit opened a large hole in the superstructure. The top of the mast disappeared over there, then the funnel was torn to pieces, and they still showed signs of life. Enemy shells whizzed into the water to the right and left, in front and behind us, toxic fountains as high as a house rose towards the sky, but we always remained unharmed. Projectiles howled back and forth more violently and more rapidly, the firing rumbled ever more furiously through the silence of the lonely sea. We were lifted up and down by the waves, and the guns thundered back and forth. Almost every shot was now hitting and devastating everything over there. Then - we almost caught our breath - the command bridge disappeared completely with a direct hit. With its destruction the ship was of course lost, since the controls were now also done for. That was not without consequences. The fire on the auxiliary cruiser stopped and a white flag was hoisted on the remains of the mast. A sigh of relief went through our boat. With our smaller ship and only two guns, we had been victorious against the larger steamer with four guns and a revolving cannon. Our joy was indescribable, especially since we were safe and unharmed and none of us had suffered the slightest injury. With a hole in the diving tank we wanted it to be over.

And what had happened over there? The effect of our hits was terrible. No sooner was the white flag hoisted than lifeboats were launched. Some of the launched boats were so badly damaged they sank as soon as the crew got in. Some of the crew quickly drowned. Skilled swimmers tried to escape over to us. But some of them, too,

---

[77] The 9500 ton Italian Navy commissioned and manned armed collier was enroute from Pensacola, FL to Italy with a cargo of Naptha for the Italian Navy. The vessel was armed with four 76mm 40 caliber cannon.

were drawn into the depths before our eyes. Only two officers remained from the Italian steamer's command staff, a lieutenant from the gun crew and the doctor. Three other officers, including the captain, had been killed by our projectiles. The last remaining lieutenant had also been wounded along with a large number of men. All were properly and carefully attended to by us. Even now most of them screamed at the doctor's treatment, who had to clean a wound here, cut one there, and sew on another. But this time we didn't care much about the moaning. Because those lying there wounded and crippled had everyone intended to destroy us without exception if possible.[78] If they hadn't succeeded, it was certainly not due to lack of trying. This time we had remained the winner, this time we were more alert, quicker and steady. But next time? Who knows how we would fare. It was war, and it always you or me.

After the wounded were bandaged, they were provided with provisions and drinking water and they could begin their journey home. Were they truly going home? In the currently prevailing weather it was a journey into the unknown. As we learned from newspapers later, only six of the 113-strong crew made it home alive.[79]

Now that we had taken care of the crew as far as we could, we began to examine the abandoned and stationary auxiliary cruiser by means of a prize crew. Our prize officer went onto the shattered deck with some of his men. Everywhere dead, lifeless, terribly mutilated corpses among the countless rubble. With open, lackluster eyes, the dead stared out of faces contorted with pain into infinite space. The command lined up the dead side by side. They

---

[78] Schultz uses the picturesque German metaphor "wenn eben möglich mit Mann und Maus zu", i.e. if possible with Man or Mouse.

[79] The commanding Capitano di Fregata Agostino Scaparro, (1869-1918) was posthumously awarded Silver Miltary Medal for Valor. Tenente di Vascello andrea Schiappacasse was posthumously awarded the Bronze Miliary Medal for Valor. The Italian Doctor referenced in the text, Tenente Medico Buosi Renato, (1889-1918) did not survive. An exact crew roster is not extant; 74 crew are known to have died. Italian sources indicate two survivors.

would be buried in seaman's fashion before the steamer was sunk,. The investigation continued. Nothing stirred, destruction and horror everywhere. But wait, what is it? What kind of rattle is that? And now a droning? The men look at each other in dismay. Then they hurry to the place from which came the unfamiliar noises, and now they heard more clearly the tearing and tugging of chains. It had to be here, behind this door. It was torn up with force and what do they see? A magnificent live young ox. This was a most a welcome booty. The submarine was immediately notified of the rare find, and the pleasant news was greeted with great joy. On the captain's orders, the cook and several other men went over to slaughter the animal in a skillful manner. After such a long time without, the very thought that we were to have a delicious meat stew and fresh meat again made our mouths water. The cook and his team were greeted with joy when they returned with the juiciest parts of the cut-up animal and set about preparing a real feast, which we had truly deserved after the victorious battle. But before we came to the table pleasures, there were one other thing to do. It was necessary to pay last honors to our fallen enemies. While we stood bareheaded, the captain made an address; briefly and clearly he pointed out the loyalty to duty which animated these men to the last moment. We honoured ourselves through the final military honours paid to them. Their sacrifice and their fighting spirit should be a model for all of us. This brief burial service on the open sea was silent and simple. Just as quietly and simply, the fallen were handed over to the waves according to seaman's custom.

Then we went to work. Two of the guns on board the "Sterope" were dismantled and brought on board because we wanted to bring them home.[80] Now at last the steamer was ready to be sunk, but at the last moment we were stopped by the appearance of a Swedish sailing vessel. This was immediately driven off, and six shots quickly informed him to stop. After a thorough search by the prize crew, however, he was able to continue on his way, as the cargo was legal.

[80] U 155 had done similar in the previous cruise, bringing home artillery pieces from two vessels.

Only now, at 1 o'clock on Tuesday, did we finally get around to sinking the "Sterope".[81]

---

[81] Korvettenkapitän Eckelmann's detailed report on the engagement follows:

After almost 3 months of warfare in the area between the Spanish-African coast and the Azores against merchant shipping for the Mediterranean, with 52,000 GRT of cargo space sunk, it was time to think about the journey home if the fuel should reach Kiel. Before that, the engines had to be thoroughly overhauled in order to be able to travel through the particularly dangerous area of the Faroe Islands, the North Sea and the Kattegat at maximum speed (In the Atlantic with normal swell and swell at about half the 10 knots at the Eckernförde Mile) and without engine failure. This overhaul took 3-4 days for each engine, during which time only one engine was ready for use, reducing the surface speed to about 5 nm and worsening the already poor maneuverability. A fight under such circumstances was not unadulterated happiness, it would also interrupt the overhaul work, which, taking into account the fuel supply, could not last longer than 8 days. I therefore moved our location to about 250 nm west of Lisbon, where I assumed there was less traffic.

Three days passed quietly, until April 17th - a Sunday - at 7:20 in the morning the announcement "Steamer in 270 degrees" broke the silence. The steamer ran smokeless and was painted gray, which is why it was discovered later than usual and was approaching very quickly. He was heading straight for us, seldom was there such a propitious opportunity for an underwater attack, which of course was greeted with elation, even if it did not fit into the program. There was just time to get the boat underwater and on course for a torpedo attack as he was in the range required for closer observation through the periscope. It looked like a commercial steamer of 8-10,000 GRT, and its speed, which I estimated at 16 nm, was unusual for such a ship. It was very embarrassing that I couldn't see any weapons. Since almost all steamers encountered so far were equipped with guns, there could be no doubt that this high-quality ship also carried guns that were well masked. Would I be allowed to carry out the torpedo attack that was in progress? A tough question. We were outside the restricted area, and according to the regulations for conducting submarine warfare, I was not allowed to shoot down an unarmed steamer here, but had to stop and examine it. In this case, that meant: putting the boat at risk in an artillery battle with very dubious prospects of success as opposed to very likely success without endangering the boat in an underwater attack. As nonsensical as the matter was, it didn't help, against my convictions I had to use the whole boat in the most dangerous manner if I wanted to ensure success, not out of necessity for the war but to satisfy a paper requirement. The bow tubes were ready: I ordered "Ready stand by", continued on the attack course, sheared into the enemy's wake and surfaced immediately, hoping the enemy lookout would least suspect a danger astern and would have directed his attention ahead. A bow shot would have ranged to 3-400 m, when I got out of the turret hatch, the distance was 1,800 m, and it had increased to 3,000 m when the first warning shot was fired from the front gun after 2 minutes at 7:52 a.m.

Until then, *U 155 had* not been seen by the steamer. After the shot he turned slightly and immediately opened a lively fire from 3 guns, the impacts of which were very soon in the immediate vicinity of the boat, so that only energetic zigzag courses and otherwise Dear God could save us from hits. I didn't expect any other development; maybe Berlin!? A single hit, even a shell splinter in the bunker, would have been enough to kill *U 155* ; even a simple loss of oil with its traces of oil would have put the implementation of the journey home into question.

That the enemy did not attack, but believed he could escape danger by fleeing, was our salvation and his ruin. However, to be honest one has to consider hardly has been previously met such a selection of outstanding gun captains and one so well trained, battle-accustomed crew, each of whom gave of himself more than any man. The steamer tried to escape with zigzag courses at the highest possible speed and fired sometimes could manage 3, at times all 4 of the guns. We were also not idle with our two 15 cm guns and were able to observe several hits on the target, while we ourselves were spared hits, despite the often threatening proximity of the impacts. After 45 minutes the distance had increased to 104 hm and 140 rounds were fired when our front gun failed (material breakage). With the aft one continued to fire alone. Since the opponent moved three times as fast as and distance had increased accordingly so far, we were very pleased when it suddenly began to decrease. Shortly afterwards the enemy stopped his gunfire, whereupon we also stopped firing (9 a.m.).

As we got closer, we realized he had stopped, had set white flags in the tops and the whole crew was floating around at the scene in 13 boats. Help for a wounded officer was called from one of the boats: we took him on board, and our ship's doctor, still barely breathing from handling munitions, looked after him as if he were one of us: both legs had been shattered by explosives. To our delight we discovered our opponent was not a merchant ship but an Italian warship - the transport cruiser *Sterope*. The Italians, 132 naval men, had fled headlong into the boats, partly in civilian clothes, partly only in undergarments, fearful that they would be treated worse in uniform. Since they had to travel 250 nm to the nearest coast in their boats (if they weren't picked up on the way), I allowed them to complete their clothing and provisions and then released them.

The inspection on board the *Sterope* showed although 12 hits had caused severe devastation, of the truly important parts only the rudder cable was shot out, damage which could have been repaired. Incidentally: 2 intact engines, speed over 15 nm, 4 good 7.5 cm rapid charging guns with modern command transmission system and measuring device, large modern FT station, plus a cargo of 5,500 t of heating oil. We already believed we had captured an ideal cruiser for trade warfare. But the dream of sending *U 155 home* with the most necessary personnel and transferring most of the crew to *Sterope* was too beautiful and, to our regret, could not be realized because the boilers could not be fired by the oil cargo. These only had facilities for coal-firing, and the coal inventory was only 200 tons. So for this beautiful ship there was only the same route as its predecessor: it was sunk. But only after recovering 2 guns and our prize crew thoroughly searching everywhere. Among the rubble in the house of cards that was the bridge valuable secret items were found among the corpses of the commander and the chief officer which was very welcome and useful for our naval war command. The war flag was also there in its beautiful, lead-weighted chest.

The lessons from this battle are not new and will also be familiar to the Italians:

Always aim the bow at danger, never give up the chase, keep steering with the propellers if the rudder fails, do not leave the ship while it is still afloat, keep secret matters, and last but not least, protect the flag from enemy hands!

Then the overhaul work resumed, the starboard diesel engine was also thoroughly overhauled. This ended the shipyard time at sea and we could go home. There was of course an uplifting mood on board. For almost a quarter of a year we were out here, far from home, on the big seas in our little boat, always dependent on ourselves. The prospect of soon having home under our feet made us all happy and eager.

Our course was now north. We hadn't had this course for two days when we came across the English steamer "Haidwood", 4150 tons.[82] He responded to our command to stop with gunfire. So he doesn't want to stop, he wants to fight. Well, may he have it. The projectiles flew back and forth again. The fight developed even more tenaciously than with the Italian. As if by a miracle, we were spared from hits this time, while our projectiles wreaked havoc on the deck of the enemy steamer. The battle continued doggedly. We could detect impact after impact, but the English toughly held out. The superstructure was progressively obliterated but he still did not give up. We threw shot after shot from our two guns. The devastation grew larger and widespread, and we were still unscathed. We had never had such a bitter fight. But our captain was calmness itself. Without any excitement, he issued his orders with constant firmness. His imperturbability affected the entire crew. All movements were carried out calmly and in a drill-like manner. Shell after shell was brought on deck. One by one they were sent over by the gunners. The guns were already blazing hot and the English still held out; all of the enemy projectiles went into the water next to us, behind us and in front of us without harming us in the least. But finally the brave Englishman reached his fate. The ship was in such a way that it could no longer remain afloat. One of its guns was simply blown overboard by a direct hit. Our fire was no longer returned by the enemy after we had fired a hundred and eighty-two rounds. The crew left the ship and tried to escape in their boats. The steamer disappeared into the waves. We followed the boats and ascertained the enemy had loaded grain and flour. During the fight

[82] The 4150 ton Harewood enroute from New York to Gibraltar with general cargo was attacked on April 13, 1918 380 miles W by S (true) from Lisbon.

an explosive shot had torn off the upper arm of the enemy captain. Two of the crew were dead.[83] The boat occupants did not ask any help from us, so we left them to their fate.[84]

After this successful fight the journey home continued. The sea had become very restless again, sufficient for the hatches to remain closed. Nevertheless, we were always on the lookout for more prey, and nothing was spared which came our way. So one day we met a convoy of twenty steamers. We submerged before we could be seen and worked our way up to the convoy underwater. The largest steamer was torpedoed and it quickly disappeared.[85] We could not make any further observations as the escort vessels searched the whole area for us immediately after the shot. So we dived as deep as possible.

A few days later, on Saturday, April 20, 1918, we sighted another convoy of several steamers with escort vessels. Submerged we made another torpedo attack on the largest steamer. We assumed we had struck our target after a terrible explosion followed.[86] We stayed under water afterwards so as not to get into a fight with one or the other escort vessel. Since there were always several of them together, it would have been more than reckless to get involved in a fight with them. We therefore stayed under water and only dared to surface again after a few hours. The convoy was finally gone.

---

[83] The Captain T. Ryan was among the dead.

[84] A summary of the survivors' report indicates the Harewood defended itself with deck guns, and used smoke bombs to conceal its movements, but within an hour and a half had been hit eight times, its captain seriously wounded and guns disabled. After which the crew abandoned ship. The Germans approached and complimented the British on their resistance, gave aid to the wounded and shared weather and sailing information to them to help get them to safety. The steamer was then boarded and scuttled, sinking stern first. Her cargo included 594,925 pounds of brass ingots. The remaining survivors safely made land after 40hours at sea.

[85] On April 16, 1918 the 7640 ton Nirpura carrying peas and lentils from Bombay to London was torpedoed while in convoy 110 miles WNW (true) from Cape Roca. The explosion tore a hole in the starboard side by the mainmast and the ship sank in three hours. All on board were rescued and landed at Lisbon, except four who were killed when the torpedo struck.

[86] No ship was sunk in this attack.

On Tuesday, April 23, 1918, we hunted our final prey. With seven shots a small enemy sailing vessel was sunk.[87] We let the crew go without stopping them.

The journey continued in a northerly direction. For a few days we did not see any vessels because we were outside of the steamship lines.

The mood on board was getting cheerful every day. We were closer to home hour by hour. We had now been cut off for almost four months and had to rely on ourselves on the wide seas. During this time we were able to receive no news, no sign of life from relatives. And we all knew conditions were by no means good when we left. There was a shortage of food then, imagine how it would be now? Surely the whole country was still hungry or starving. We ourselves did not suffer from hunger directly on board. But swallowing canned food day in and day out was no longer bearable. We all longed to have fresh vegetables, potatoes, and especially fresh bread again. Some of the crew had already contracted mouth rot as a result of the canned food. It was high time we came ashore again. In addition, close coexistence in the small space of the boat became almost unbearable for everyone. It was not possible for an individual to indulge in his own thoughts for even a short time. There were always several around you, or you were in someone else's way yourself. It was imperative for a submarine crew to come ashore after a not too long interval, temporarily disband, and then meet again refreshed.

All conversations on board naturally now revolved around the expected stay on land and of leave. Everyone had already prepared for this. At least it was possible for us to fish one or the other useful thing from sunken ships, especially food, even if it was only canned food. We wanted to help our starving relatives.

[87] The 54 ton wood ketch Frances was carrying coal from Swansea to St. Brieuc was stopped 6 miles S from the Lizard. There were two fatalities, one due to enemy action and one due to exposure. The number of survivors is unknown.

The strength that had disappeared through long periods in the boat's bad air now revived in anticipation of this holiday. Everyone was at his post with tense nerves so we could get through the worst part of the return journey safe and sound. U 155 pushed completely blind into the danger zone. The tension was so strong that with every unusual shock one immediately assumed we had stumbled upon one of the mines which float en masse in the sea. There was good reason for this fear, however, because very often we could only avoid this dangerous thing at the very last moment, especially at night, as it was particularly difficult to identify floating mines in the dark. We had to drive very slowly because the mines could only be seen from a very short distance. Only at the last moment was there the possibility of being able to avoid danger. The watch on the tower had to be extremely careful. We engine room personnel didn't have it all that bad with our engines, especially since we knew we could rely on our comrades above. Each of them did their best to contribute to the home journey's success after we had successfully survived so many dangers and battles.

Nobody was allowed on deck. All hatches remained closed as we had to be constantly ready to dive. We, the engine room staff, were only allowed to alternately smoke a cigarette in the tower or with our heads above the tower hatch. With such precautionary measures, the boat advanced slowly but surely, and it was only due to this caution that we owed our luck. For example, some submarines were driven with less care, such that they hit a floating mine and their crews found death in the waves. We experienced submariners knew how extremely difficult it was to pass through the English restricted area, and we adjusted accordingly.

When we visited the tower, of course, we always asked about the status of the voyage and how far it was to the guard ship "Hannover". Finally one evening it was said we would probably reach it tomorrow morning at 6 o'clock. So one more night and then we'll have made it, then we'll be safe. Was this the last trip, or do we have to go out again? We didn't know the answer to this. But we now knew one thing: this war was lost for Germany. While everyone was starving in Germany and not a single country came into question as a supplier of food, the whole world was open to our opponents. We had seen it: countless steamers supplied the enemy armies with everything they needed, with soldiers, ammunition and food. And Germany still had to give up the little it had to its allies. No and again no, Germany could no longer win this war in spite of her heroic army, despite all the sacrifices of the people.

Germany could not replace the currently lost submarines; the few still sailing the seas could not endanger the enemy with resounding success. Too many fully loaded vessels, large and small, reached their destination. We couldn't stop it. Lately the conversations on board have often were like this. We made no secret of these convictions, even if we still had the indomitable will to harm the enemy as much as possible. The enthusiasm that existed at the beginning had long since given way to calm objectivity, coupled with a dogged will to fight in order to destroy at least as many enemy ships as was possible.

We were allowed on deck again the next day having successfully made the journey home. Hatches opened and air let into the boat. Happily, the off-watch personnel went up there. A few more minutes and we saw the guard ship "Hannover" in the distance. We stopped at it and tied our boat to it. Our captain went up to report, and after a while we were able to continue our voyage to Kiel through home waters. On Saturday, May 4th, at 10.30 in the evening, we entered the port of Kiel. We docked at the shipyard on Sunday night. Married couples living in Kiel were allowed to visit their families the same night. I did not make use of this privilege myself, but stayed on board so as not to disturb my relatives during the night. However, early on Sunday morning, I hurried to my family with a cloth sack full of groceries. My sudden appearance caused an indescribable joy. It was frequently reported by submarine personnel that a number of submarines had been reported missing, including U 155. One can imagine my wife went through anxious days and nights in response to this news, and also an overwhelming happiness when I suddenly came back safe and in good spirits.

The day of reunion came to an end and it was time to go back on board. The duty wasn't too difficult. In addition, I received a few days of leave and used it to visit my parents in the Rhineland. I knew from my last visit that the old people were suffering terribly from lack of food. So I packed half of the hoarded canned food and took them with me. Because it was impossible for a leave taker to get full on the food obtained with great difficulty from the grocery store. And we submariners were used to plenty of food, including even now in the shipyard. Without a reasonably adequate diet, it would not have been possible to meet the sometimes monstrous demands of the submarine service. The so-called rest periods were often only on paper. All too often, as a result of alarms, attack preparations, underwater voyages, twelve, eighteen or even twenty-four hours of uninterrupted duty had to be performed. And one always had to be tense with attention in order to comply with every sudden command as quickly as possible. It was precisely this stress on the nerves that made submarine service so difficult. Certainly the other inevitable inconveniences were also difficult to bear, such as living in narrow spaces, bad air, heat, etc. But all of this was insignificant compared to the concentrated attention with which one had to work on the engines. For the slightest mistake, the slightest negligence could have devastating consequences. And without adequate nutrition, it would have been impossible to meet such demands.

The leave with my parents was again a torture for me. I could hardly watch the starvation and the apportioning of the little there was so that everyone got exactly their ration. I would have loved to turn around again, but I didn't want to do that to my parents. They were so happy to have me back home with them. I had to recount in detail all my experiences, including those from the journey to America. Neighbors and friends also wanted to hear about my experiences, and so the leave passed better than I had feared. Here and there was lunch or dinner for me at one or the other, so my parents were relieved a little. Still, I was happy when leave was over. On board we didn't have to worry about food issues.

Duty in Kiel went its usual way. Crew quarters were on a barracks ship while the boat was in dock. After the end of work, at 5 p.m., the married men were allowed to go ashore to their home. It was a calm, orderly life in which excited and tense nerves gradually recovered and strengthened. In addition, the food was so good and plentiful that we were mostly able to bring some bread or other food home for our women. Often we thought of the huge amounts of food we had sunk out there. How necessary, how bitterly needed these quantities would have been here. Here in our country there was famine, and out there we had to destroy everything that was missing here, with which countless hunger could have been satiated. The quantities that fell victim during the activities of our recent cruise should also be mentioned. We had sunk: nine steamers, seven sailing ships, plus the Italian auxiliary cruiser "Sterope", a total of 48,247 gross tons. The cargoes sunk with them consisted of 9,700 tons of grain, 7,500 tons of flour, 5,000 tons of rice, 6,000 tons of brass and wire, 11,000 tons of cotton, 470 tons of salt, 300 tons of salted hides, 450 tons of timber (45 tons of brass we were able to bring home with us bring).

VI. In a U-Cruiser To America

In July 1918, work was carried out at full speed on completing the boat's repair. Not only regarding us, but also at all shipyards, in order to send as many submarines as possible against the enemy, because they were the only effective naval weapons. We too sped up the overhaul work so we could proceed against the enemy again as quickly as possible. We were ready on August 10th. Another short leave on August 11th to say goodbye again and this farewell was harder than before. On land, the Fortunes of War had already taken place and this turn was bound to make our enemies on sea more confident and more aggressive. Not to mention that their anti-submarine defenses had made further progress in recent months. At our departure in January the prospects of a successful return had been much better than now. And yet there was no such thing as discouragement among our crew. The only thing we wanted was to sell our lives as dearly as possible. A grim cold power had taken over us.

On August 12th one last attempt at diving and trimming, and then after a short journey we docked again at the watch ship "Hannover". At 10.30 in the evening, a torpedo boat towed us, and - two more destroyers ahead for greater security - we drove through Swedish waters at the fastest possible speed. In the morning at 5 o'clock an airplane appeared which was also watching the waters for our safety. At noon the escort boats and the plane left us and we were again dependent on ourselves.

Our new commander, Lieutenant-Commander Studt[88], announced to us on deck that we had the assignment to travel to America, lay

[88] Korvettenkapitän Ferdinand Studt (1879-1958). U 155 was his first submarine command.

mines and cut cables. So America again! But very different from the previous year. Now we will probably not be received and entertained with boisterous enthusiasm. But what did that matter? It was war and we were doing our duty.

There was no possibility to think about the changing times because the trip through the Sound and Kattegat was a very uncomfortable thing. The many mine fields required everyone's attention. It was late at night as we drove around Skagen. At 3 a.m. in the morning we saw the Skagen beacon. There we were noticed by an enemy cruiser, which is why we quickly dived to within forty meters. The course was then changed under water, and the journey continued closely by the English mine barrier. Not until noon did we find our way through the middle of the mine field, and during that time came hours of extreme danger, hours of extreme tension, because every second we expected a mine to explode and a mighty flood of water rush into the boat dragging us all down to the ocean floor. No fight with a visible danger, with a tangible enemy is as grueling as this hours long voyage on the verge of death. We certainly had dangerous skirmishes with superior opponents on our last cruise. But we never experienced the heart-pounding stress of driving for hours through this treacherous minefield.

Just as everything comes to an end, so did the horrific death ride through English mines. As if freed from the weight of hundreds of pounds, everyone breathed a sigh of relief when we passed through successfully late in the evening. And immediately we were able to enjoy ourselves by stopping a steamer.

Despite the exhausting day, we only had a little sleep, because suddenly at 3 a.m. there was an alarm for a quick dive because a steamer in sight was accompanied by a destroyer. Another steamer was sighted at 6 a.m., this time without the uncomfortable company of a destroyer. First of all we dive. Submerged we worked our way closer to the steamer and reappeared some distance away. Various warning shots quickly taught him he had to stop. Unfortunately it was a wasted effort. Our prize crew found the cargo was in perfect condition, so we had to let the vessel continue.

Now came a few boring and uneventful days. On August 23rd we drove on the open ocean again, but unfortunately in a stormy sea which even threw our heavy U-cruiser back and forth. We could therefore only drive with one engine only and made slow progress. But this inconvenience only lasted a few days. The weather soon got better and the sea calmer again. We quickly came back to life on board. All the off watch went up to recover from the grueling work during the stormy days. Suddenly the tower watch called into the calm: "Several columns of water can be seen on the horizon, apparently a submarine is fighting with a steamer." At this report, we quickly disappeared to diving stations. The boat went under water in a flash and headed for the reported water columns. We prepared for a bitter struggle. If one of our submarines was in danger, we had to come to the rescue as soon as possible, before it was too late. We had only one wish to get to the scene of the fight as soon as possible. So we hurried there at our top underwater speed. Through the periscope it was seen the water spouts were approaching our boat with a speed never before experienced. We were faced with a riddle, the distance between us and the water columns was noticeably decreasing, and then we were very surprised to find the water spout came from a pack of whales. Immense laughter went through the whole boat after this discovery. We quickly surfaced to better see the rare spectacle. Everyone who was off watch hurried up on deck. We were now able to observe the animals up close. Each of them released a high jet of water about every two minutes. The huge animals romped about in the water with marvelous agility. We could not determine how many there were, since the animals dived up and down very quickly. Among them were several examples of almost pre-antediluvian size. The whales passed us fairly quickly. We watched the water spouts for a long time until they disappeared on the horizon. We had seen a picture of nature's rare beauty. For a long time afterwards we were amused by the prank the animals had played on us, and many a tower lookout was admonished to take good care for whales.

On the last day of August, at 3.45 p.m., we stopped the Portuguese three-master "Gamo". The crew climbed into fourteen small boats that could only fit two men. In one of the boats was the captain with his eighteen-year-old son. He came rowing towards us and pleaded with our captain to take at least this boy of his, his only child, on board. Since they would unlikely ever reach land in their boats, his wife would at least have a breadwinner. But our captain couldn't make an exception. He had to be tough on the enemy. The sailing vessel was sunk. The small boats, which were probably not able to cope with the slightest swell, remained behind on the troubled sea.[89]

While the sailing vessel was still sinking, we saw large clouds of smoke rising on the horizon. Soon we recognized three large express steamers with four funnels each. They too noticed our boat and shortly afterwards opened fire. Of course, we couldn't get involved in such a superior force. So we quickly sought a safe depth.

The next day we made an attack on the Norwegian steamer "Stortind". A torpedo could not be used because of the rough seas. The swaying of the boat, even under water, was so great it could hardly be held. So we surfaced and attacked the ship with artillery fire. After five shots it disappeared, and with it a load of 2,500 tons of iron goods, rice and sugar. [90]

Alarm bells rang again at 10 p.m. We were pursued by a steamer, apparently an auxiliary cruiser. So quick dive and down to a depth of thirty meters, where we were safe from the enemy. We couldn't get involved in a serious fight and risk being sunk.[91] We still had a

---

[89] The 343 ton fishing schooner Gamo was outbound from the Tagus in Newfoundland water for the fishing season. Master João Fernandes Agualusa commanding. The vessel was stopped approximately 370 miles from Flores Island, Azores. The Portuguese captain reputedly responded to the German captain's inquiry of where he would go with the dories, "I'm going to the bottom of the sea, because where else can I go in this nutshell?" The dories were each 13' length x 5' width. 34 of the 39 crew were saved after nine days at sea and traveling approximately 470 miles in which 5 dories were swamped and lost.

[90] The 2562 ton Stortind was carrying general cargo from Norfolk to La Pallice. Mastern Christen Reed Erichsen commanding. The vessel was stopped on September 2nd 1918. The Master's boat was picked up after 8 days by the British S.S. Tudor Prince. The Mate's boat with 14 men arrived September 6th at Flores Island, Azores.

[91] A reference to a long range gun duel with the armed tanker USS Frank H. Buck, the latter claiming to have sunk U 155. The engagement commencing at 14000 yards

lot to do over there. We were now pretty much in the middle between Europe and America. The journey so far had been very strenuous - but the most difficult and dangerous part was just ahead of us. Laying mines and cutting cables off the enemy coast thousands of kilometers from home were after all these tasks of a special nature. So we sometimes had to avoid a fight were we would otherwise have liked to fight. But what was possible was sunk. i.e. the little steamer[92] "Sophia" on September[93] 7th. Its crew were fishermen. Immediately afterwards we chased a steamer, but it escaped us because it was able to develop superior speed.[94]

In the meantime we had come near the American coast and from now on it was important to proceed with extreme caution again because one small incident could be enough to thwart our plan. In addition to this, there was a heavy fog that obscured visibility. There was nothing left to do but to continue the journey slowly with the boat's hatch closed - ready every second for a quick dive. After a few days the fog finally went away, but the sea was rough. It had gone from bad to worse, because a heavy swell, present near the Newfoundland Islands, was a very uncomfortable experience for us. And just then - on September 12th, we sighted an American cruiser for the first time. Like an electric spark it went through the whole crew, no matter how much they were drained by the exertions of the last few days. And immediately our captain made the necessary arrangements to launch a torpedo attack. But it was in vain. The enemy was too fast. Before we could get close to him with the correct shot setup, he had been able to gain distance in the strong swell.

---

reducing to 10,000 yards. The tanker's aggressive handling and 6" stern gun undoubtedly providing the impression of being an AMC.

[92] A recollection error on the part of Schultz as Portuguese sources indicate the vessel is a 162 ton un-motorized wood fishing schooner. Vessel was loaded returning to Lisbon after catching cod off the Newfoundland Banks. There were no fatalities.

[93] Typographical error in original indicated August.

[94] The British S. S. Monmouth, 4,078 tons escaped. U 155's engines may have been underperforming as the 14 year old Monmouth was rated for 11 knots vs. U 155's 12 knots. Seas may have affected the submarine as well.

Instead in the evening a Portuguese ran into us. It would not escape us. It was the steamer "Leixoes", 3245 tons. A torpedo hit sank him within a few minutes. The crew, who quickly got into their good boats, probably did not expect to be attacked so close to the American coast by a German submarine. When we approached them, they asked us for drinking water, but we could not give it to them because we ourselves were deficient in it.[95]

Despite the proximity of the American coast, a new enemy was attacked at dawn the next day. A torpedo was fired after difficult attack preparations due to the stormy weather. But on board the steamer they were paying close attention and identified the torpedo's wake in good time, quickly changed course and it passed by. We surface. The steamer was immediately fired at with our guns, whereupon we received counter fire from two cannons. A lively shootout developed. Various hits from us could be observed. But the steamer tried briskly - always firing at us - to escape. We didn't give up, shot after shot was sent after the pursued. He kept sending out distress calls. The battle went on at length. We had already fired one hundred and eight rounds when we suddenly had to break off the fight and the pursuit. An enemy dud ripped through a rivet, creating a huge dent in the pressure hull and damaging various oil lines.[96] Our boat was unable to dive due to this damage.

---

[95] The 3345 ton Leixões was in ballast from Hull to Boston. Master Joaquin F. Sucena commanding. The ship was 200 miles from the Newfoundland cost when attacked on September 12, 1918 and sank in 15 minutes. One man lost his life in the torpedo explosion and two more died of exposure among the four lifeboats the crew of about 50 men got into.

[96] It was the British flagged 4391 ton Newby Hall enroute from Milford Haven New York, Master F. O. Seaborne commanding and his report follows:

The Newby Hall left Barry on August 30, 1918, and left Milford Haven on September 1, 1918, in convoy, bound for New York. On September 3 at 6 p. m. we were ordered to detach convoy and proceed to destination according to secret orders. All went well until Friday, September 13, at 9.52 a. m., in latitude 42° 18' N., longitude 58° 22' W. We sighted a torpedo coming toward us from three points on the port bow. We immediately put helm hard starboard. The torpedo missed by about 6 feet, passing our bows from an angle over to the starboard side. 1 then steered parallel to track of the torpedo, and at 9.56 a. m. (approximate) I altered to SSW., thereby bringing position of the submarine astern and instructing engineers to give all speed possible.

At 10 a. m. we sighted submarine coming to surface bearing north; I then kept him astern steering an irregular zigzag course (about 3,000 yards).

Without thinking too much about what could and would now happen, we immediately set about repairing the damage suffered. Fortunately, the SOS calls sent by the steamer were without immediate succor. It there could have been dire consequences for us if help had come. In an emergency we could have gone under

---

We saw their gun crews coming out of the conning tower and manning the two guns. They commenced firing immediately and we replied with our gun. He was then steering in a westerly direction and going at a moderate speed, and to keep him astern we had to gradually alter course to the eastward.

The enemy was firing rapidly with both guns. The forward one appeared to be of larger caliber than the other. After about 50 rounds with our gun a direct hit, smoke, flash, explosion put his forward and largest gun out of action. When the smoke cleared we found this gun had tilted over to an angle of about 30° and no gun crew was to be seen. Soon after we scored another hit on his forward end, the fore part of the forward gun causing an explosion and a volume of smoke. After that we scored another hit on his after end, about 20 feet aft the after gun. He then reduced his speed and seemed to be under difficulties, but continued firing with his after and smaller gun.

At about 13,000 yards we outranged him and he ceased firing. He headed toward us and appeared as if he was chasing us, but in a few minutes he was broadside on to us again and stopped and suddenly disappeared. Action lasted from 10 a. m. to 11.20 a. m., when firing ceased. Enemy commenced and ceased firing first.

During the action the enemy used shrapnel and high explosive shells, but none hit the ship except shrapnel. None of the crew was injured. Two boats on the port side were riddled and one plate amidship on the port side dented and badly cut in numerous places by shrapnel or burst shell. All the crew behaved splendidly throughout the action, and special praise is due to the gun layer for his coolness and the masterful way he handled the gun, and also the gun's crew who worked very hard; also the officers and engineers who assisted in passing up the ammunition.

Distress signals were sent out by wireless for assistance and answered immediately by United States patrol, who endeavored to come to our assistance but as the weather became thick and rainy we saw nothing of him. All secret books and codes were thrown overboard during the action.

The submarine was about 250 to 300 feet long; black in color; cutter bow; flush deck; gradually sloping into the water at the stern. There was a platform about 3 feet high amidships and 40 to 50 feet in length; in the center of the platform was the conning tower about 10 feet long at the base and 4 or 5 feet high. The after vertical portion of the conning tower was not parallel with the forward portion as the conning tower was smaller at the top than at the bottom. The forward gun was about 20 to 30 feet from the forward part of the platform. A mast or periscope extended from the forward portion of the platform and another was situated on the after portion. They were the same height as the conning tower.

water, but so much would have leaked from the damaged oil lines that the resulting oil trail on the surface of the water would inevitably have given us away. We worked hard to restore the ability to dive to some extent. However, the next time we went underwater we found the pressure hull at the damaged area was completely watertight.

No further attacks were made in the following days. We literally crept to the port of Halifax, mostly underwater, so our operation could not be thwarted at the last minute due to the many ships cruising around here. The mines to be laid were prepared for laying. Surface minelaying was not possible because the sea was too restless.

On the night of September 17th, we entered the port of Halifax. Before that we prepared six mines for laying, which we quietly discharged overboard. We had to work with unparalleled caution and circumspection so we would not be betrayed by creating noise or waves. And despite all that, it almost went wrong. We came pretty close to an American steamer lying in the harbor when suddenly a man with a lantern appeared on deck and tried to illuminate the water. Apparently the suddenly awakened suspicion calmed down due to all of our compartments being dimmed and an absolute silence prevailing, then the man with the lantern disappeared and we went unnoticed.

That wasn't the only danger in this work. The slightest carelessness could cause a mine to explode and we would all have blown up. This arduous and dangerous work lasted two nights. You can imagine how much we breathed a sigh of relief when it was finally done and we were able to move away from this dangerous area.[97]

Now came the second part of our task: cutting through the cables. Of course, that was not possible on bright days. We would inevitably have been discovered. So it would be at night. A seemingly

---

[97] U155 laid a series of mines 6 to 9 miles off the coast between Betty Island to the SW of Chebucto Head, the Sambro Lightship, and Sambro Island.

impossible operation; yet we had to get down to it and achieve success under any conditions. On the night of September 20th to 21st, we drove to the place where the cables should be. But the devil seemed to have a hand in it. Just in this little spot of the wide sea, three steamers lay still and motionless like sentries. What did they want here? We crept closer, and then we saw: they were steamers waiting for the next day to enter harbor. We did not carry out any productive activities that evening. "Quietly" we crept out of these inhospitable waters and rested at a safe depth during the day. On the afternoon of September 21st a renewed stalking began. This time we were luckier. The three guard steamers were gone and now we could begin our work. But oh bad luck, we had the same misfortune as our time off Lisbon, and this time at the start of work. The cable cutter snapped off and the large cable cutter disappeared to the bottom of the sea. We had to beat the bushes[98] to get the second cable cutter ready. But unfortunately it didn't hold up much better than the first either. Again after a short period of work it was lying on the seabed. With that we were at the end of our tether and at the end of our cutting activity. It was not without noticeable relief that we left the dangerous place. We were used to all sorts of things, but laying mines and cutting cables in an American port was a job that made the highest demands on the nerves of the crew and the maneuverability of the boat.[99]

The rest on the following days was well deserved and the encounter with an American steamer was just a pleasant change. "Kingsfisher", that was the name of the steamer, was sent to the bottom of the sea.[100]

---

[98] Schultz's own expression.

[99] U 155 was in the area of Sable Island and failed to cut any cables linking Canada to the UK.

[100] The 353 ton fishing trawler Kingfisher was out from Boston when stopped on September 20th 1918, 85 miles from Halifax. There no fatalities among the crew of 27 men. Master Riley commanding and reported regarding the Germans:

"The commanding officer was about 6 feet tall, thin, ruddy complexion, sandy hair, wearing short side whiskers, and appeared to be between 25 and 30 years of age. Another officer of the submarine was about 5 feet 8 inches tall, light complexion, clean shaven, and weighed about 180 pounds. About 53 men were

Unfortunately, it was to remain the last victim for the next few days. From September 22nd to 26th we roamed to and fro near the coast seeking in vain for new victims in the otherwise busy area. It was not until September 26, at 5 a.m., that two sailing vessels came into view. Since we couldn't control both of them at the same time, we hoisted the American flag and drove through between the two of them with a cold smile.

An hour later we saw a lone large steamer. We immediately dived and tried to attack underwater. But since we couldn't get close enough due to the enemy's speed, we went back to the surface to attack with the guns. Upon surfacing, our commander discovered two submarine hunters near the steamer.[101] Of course we couldn't get involved with these dangerous fellows. We quickly went under water again, where we stayed until 2 p.m. After surfacing, we drove away from this poor pasture and further out into the sea. And now we kept a course to the east, that is, the course to home. Finally, finally!

As pleasant as this was, the next few days were just as unpleasant. It was not until September 29th that we caught another opponent, who we couldn't attack under water because of his speed. We then surfaced and attacked the enemy with gunfire. But he apparently knew very well what tactics which one could used to counter submarine attack. He zigzagged at top speed, such that all our shots missed, and kept radioing for help. After a short time we had already fired fifty-four rounds at the steamer without getting a hit, when our mortal enemy, a destroyer, summoned by the distress calls, rushed over. We quick dived. Eight shells, which we could not bring back inside the boat, were unceremoniously thrown overboard. We kept quiet for a few hours. As we surfaced, we saw the destroyer was still looking for us. Our boat then went back to a

counted on the deck of the submarine, wearing various kinds of uniforms. The crew were all young men and apparently healthy and contented."

[101] The vessel's "Allo" was received on the U.S. mainland and the ship was not identified further.

safe depth and did not go up again until 11.45 a.m. Now at last the pursuer was gone.[102]

The next few days brought us new disappointments. On September 30th, we chased a steamer for several hours, but it turned away and evaded the chase with superior speed. The next day we sighted three steamers, but again our attempt to get to them was in vain. They were traveling so fast we couldn't catch up with them. On October 2nd things were better at first. Another steamer was chased and shot at when we got close enough. But after the first shot the steamer, with surprising impudence, changed course and sped towards us and tried to ram us without further ado. We could only avoid this attack by quick diving; ten shells and ten powder bags had to be thrown overboard.[103]

Finally another American warship came into view. Here was an opportunity to make a big hit after so many failures. With all engines at emergency speed we pursued the enemy in order to intercept him properly. After a long hunt, the moment had finally come. The command sounded, and at the same moment a well aimed torpedo left the tube. According to our calculation, the shot had to hit. But after a short time we realized it had failed: the launched torpedo had not kept its set course. We assumed the torpedo mechanism itself had somehow failed. We couldn't get a second shot because the warship was too far away and could no longer be overtaken. We were gripped by bitter annoyance when we saw the warship happily steaming along at dawn on October 3rd.

But the fortunes of war were favorable to us. About two hours later, the sharp eyes of the lookout spotted another column of smoke,

---

[102] The 2246 ton British S.S. Reginolite was the vessel. The submarine opened fire at a distance of about 5000 yards and continued firing for 45 minutes without result. The Reginolite returned the fire while steering a zigzag course at her best speed. The October 1917 completed vessel was rated at 8 knots. The submarine was outdistanced and gave up the chase and submerged.

[103] The 9071 ton British steamer Nevasa on a voyage from Glasgow to New York, W.H. Walton commanding, reported the October 2nd submarine contact differently. The submarine was about 8 miles away when first seen attempting to overtake the steamer without success; then swung broadside to, fired one shot which fell far short and abandoned the chase. The 14knot Nevasa quickly outpacing the submarine.

which we quickly headed towards. An attack was immediately launched. "Adolfo Treves" - that was the name of the Italian steamer of 3838 tons - was not alert and nimble enough to avoid the impending disaster. The torpedo fired at him turned out to be a good hit. We surfaced to observe the effect of the hit and possibly to finish him off with a few cannon shots. But as we got closer, we noticed in the distance the thick plume of smoke from another ship which was heading rapidly for the "Adolfo Treves". Since it was most likely an American destroyer hurrying to help, we didn't get any closer to the struck steamer. We just watched the crew leave the ship and get into their boats. Before the assisting ship got any closer, our gun crew fired ten more rounds at the Italian, from which bright flames then burst. We no longer waited to see the fate of the steamer, but instead disappeared to a safe depth before the approaching destroyer.[104]

The last few days made it very clear to us how difficult our struggle had become. Not because of the dangers but because all vessels now steamed with the greatest caution and at high speed. That is why very often the enemy escaped us if he noticed us in time. For example, we missed enemy steamers on October 4th and 5th, because they noticed us too early in the fairly calm sea. The loot these days consisted essentially of a small American sailor named "Industrial."[105]

In addition to the displeasure about the low successes, there was another circumstance pushing the mood to a low point. It was the increasing heat the further we got into the Gulf Stream, which we had to cross on the course we had chosen. Besides the heat, the sea was very choppy. Waves as high as a house tossed the boat up and down. In addition, all the hatches had to remain closed due to tropical downpours. Therefore it became oppressively humid and almost unbearable to stay inside the boat. You would think you

---

[104] The 3838 ton Alberto Treves was voyaging from Cartagena to New York. There were 21 dead among the crew of 34.

[105] The 330 ton wood schooner British flagged and Canadian crewed Industrial was on a voyage from Turks to Nova Scotia with a cargo of salted cod. There were no casualties.

were suffocating in the hot and humid air. Most of the crew walked around bare-chested, yet sweat poured down from them. There was no longer any thought of sleeping. The heat and hideous smell of oil, plus the boat swaying, made it impossible to doze for even a quarter of an hour. You laid down in your bunk, put your knees against the side of the ship, your feet against the outer wall of the bunk and at least tried to let your limbs rest a little. You were happy when you weren't thrown out. Because of this, many preferred to lie flat on the iron deck in the control room in order to find some cooling here. You could hardly stay in the berthing section, as everything was mixed up there. Then you began your duty watch exhausted and bathed in sweat. Anyone can imagine in what condition you were at the end of it. This hellish life lasted for four days and four nights.[106]

We passed the zone of terror on October 9th. The weather also cleared up a bit and the sea was calm again. And - we were allowed on deck again. Quickly the hatches opened and out we went into the fresh air. Cigarettes were lit with ravenous hunger and greed; having to do without them while being imprisoned on such days was the worst evil. That is why we were most happy when allowed back on deck. In the fresh air and with happy laughter, the torments of the last few days were soon forgotten. Our commander also ordered all engines turned off and the boat remained still in the fine weather. Everyone would have the opportunity to recharge their tired nerves and tired body after the stressful days. We stayed in this wonderful calm for hours.

The smooth sea and the wonderful clear weather truly invited swimming. The captain happily gave permission, as he knew this was one of the best ways to give limp bodies new freshness. We quickly undressed and went headfirst into the water. A rope ladder attached to the ship's side was used to climb back up on deck, only to jump into the water again after a short time. A seaside resort has

---

[106] Studt commented in his report how the air conditioning system which was built for a merchant crew of twenty-nine men was unable to meet the needs of the wartime seventy-two man crew.

never seemed so refreshing to us. We couldn't have enough of it. But suddenly we were disturbed in a very uncomfortable way. The man on watch in the tower suddenly shouted excitedly: "Get out of the water, there is a shark." Terrified, we hurried out of the water. The last one had just pulled up with the rope ladder when the beast was already shooting up. We were shocked when we thought about what would have happened if the lookout hadn't seen the monster in time. The shark circled pretty close to our boat, and after a short time three more were added. For hours without a break, the four of them made their way around the boat. Then it occurred to one of the people that one should try to catch one of the animals. Immediately an appropriate line was cast out, but the beasts were not in the least bothered by it. Then an attempt was made to kill one of the animals with a pistol. But the shots had no effect on what they were doing. They circled the boat quietly and at a steady pace, estimated at a depth of about two meters. Bathing in the open sea was over since the animals apparently did not want to leave us for the time being.

The quiet and the refreshing bath quickly made the entire crew fit for action again, and we thirsted for new experiences. Our thirst for action should soon find an opponent. On October 12th we came across an armed steamer, with which the fight was immediately started. The enemy did not stand for long, however, and soon fled. Of course we constantly fired over to him. One hundred and sixty-six rounds were fired at the fleeing vessel. Four hits could be clearly observed. But the steamer had a faster speed than us and in the end escaped.[107]

---

[107] The 7409 ton USACT (United States Army Chartered Transport) Amphion was returning to New York from Bordeaux in ballast. Lieutenant-Commander H.H. Norton commanding. The Captain reported the submarine's second shot carried away the steamer's wireless. There ensued a gunfire exchange of 1 hour 20 minutes in duration, the submarine firing some 200 shells and the Amphion 72. The Amphion was repeatedly hit, lifeboats riddled, superstructure damaged, but she gradually drew off and the U-boat abandoned the chase. Two men were mortally wounded and a number of others were seriously injured by shrapnel. Chief Boatswain Mate

In pursuit of this steamer, we reached the southernmost point of our voyage. Now we turned from the almost southerly direction we had kept in the last few weeks and headed north-east, i.e. the course towards home.

And it almost seemed as if the change in the travel direction we all eagerly awaited was about to be our last hour. On Monday, October 15th, we experienced one of our most terrible days. At 4.10 a.m., a test dive was carried out as usual. During the underwater trip, the rear hydroplane suddenly failed. As a result, the boat was pushed deep into the water. The stern sank so much we could hardly keep our feet within the boat. Considerable agitation seized the crew. It must not happen that we all are going to die because of damage to the boat. So everyone went to great lengths to bring the boat back to a normal position. It absolutely had to be made steerable again, or we would be lost. Finally, after the last resort was implemented - putting compressed air into the diving tanks - we managed to get up. Of course, once on the surface we immediately started to make the damaged hydroplane usable again. Work was not finished until 11.30 a.m., and we were able to go to sleep with a reassuring feeling of newfound security.

Two days later occurred the last sinking of an enemy steamer. We sighted a convoy and immediately maneuvered our way up to it and fired a torpedo at the steamer "Lucia", 6,700 tons. We closely observed the torpedo hit the "Lucia". Submerged we waited for the further effect of our hit. "Lucia" kept sending out calls for help and apparently managed to still stay afloat. Not until the next day did we hear her radio message, which read: "We're getting into the boats because our ship is sinking." The crew was picked up by another steamer a little later. [108]

William Henry Schmitt of the Armed Guard was awarded the Navy Cross and Army 2nd Lieutenant Henry John McGrath the Distinguished Service Cross for heroism.

[108] The 6744 ton Lucia was voyaging from New York to Marseilles. Master C.F. Leary's detailed report follows:

"At 5.30 p. m. on October 17,1918, the U.S A.C.T. *Lucia,* bound for an European port in convoy (without escort), was torpedoed in the engine-room on port side, killing four men. The position at the time was latitude 38° 05' N., longitude 50° 50' W.

There was no sign of any submarine to be seen. I had just finished my supper and was going on the bridge. The commander of the armed guard, who was on top of the chart house several minutes before I got on the bridge, said he saw the wake of the torpedo about 100 yards off, and had guns trained in that direction. There was an efficient lookout kept at all times. Our speed at that time was 10 1/2 knots. The S. S. *Hawaiian,* our guide ship, was about two or three points on our port bow, about 1,000 yards distant. We immediately sent out S O S calls and our position, and semaphored the guide ship *Hawaiian,* to broadcast same. We sent out S O S calls from time to time on auxiliary set, until five minutes before leaving ship.

As we fully expected the U-boat to return and send another torpedo into us or shell us, I decided after a consultation with the commander of the armed guard and officers, to lower three boats and put civilian crew afloat and hold on to bow and stern of the ship. As our No. 4 lifeboat was destroyed by the explosion, and her complement was 22 persons, four of whom were killed in the engine room, 18 persons were left to divide up in the remaining boats. There three boats hung by ropes from the stern of the ship all night until morning of the 18th of October. I had civilian crew come on board and get some food and return to their boats. Our after decks were awash at this time (1 a. m.). Our engine room and fireroom and No. 4 and No. 5 hold were full of water, No. 6 was one-half full and gradually filling and deck load breaking loose. The wind and sea was increasing, and seas breaking over our after decks. About 10 a. m. we put No. 2 boat afloat, bringing her on lee with three men and stood by, as ship was slowly settling down.

About this time I heard a noise below, which I believed was the carrying away of the bulkhead between the fireroom and No. 3 hold as the ship settled more by the head after this, which I believed was due to the flooding of No. 3 hold. At 1.30 p. m. we launched No. 5 boat, which was on starboard quarter, and hung her off to leeward, standing by. We also launched after life raft and forward life raft. The after gun crew had to abandon their gun as the sea was breaking over gun platform and they were ordered to their boats immediately. At 2.20 p. m. I left for ship in No. 2 boat with the commander of the armed guard and forward gun crew and three others of the civilian crew. No. 6 hatch was broken in by deck load and filled. Stern settled down, with after decks under water. All boats hung at leeward of ship until 3.15 p. m. when her stern sank until ship was perpendicular with her upper bridge just underneath and sank in that position. At 3.20 p. m. she disappeared beneath the waves. After the ship sank, all boats hauled away clear of floating wreckage and laid to sea anchors. A very heavy sea was running now and our boats were taking considerable water on board and we had to keep bailing continually. At 9.26 p. m. October 19, we sighted the U. S. S. *Fairfax* coming to our rescue. At 10 p. m. we in No. 2 boat were all landed safely on the destroyer, and at about 12.05 a. m. the last boat was landed on board the destroyer without any mishaps.

If the *Fairfax* had not arrived at the time it did, I do not think our boats would have weathered the heavy seas, as they were all overloaded. I and my officers and crew think that great credit is due to the commanding officer and crew of the *Fairfax* in the skillful way he handled his ship, in the heavy seas, and effecting the rescue of our crew without any mishaps.

All confidential books, codes, and papers were dropped overboard immediately after we were hit, in presence of armed guard commander and ship's officers. I

Once again we made an attack on a convoy consisting of two cruisers and a large steamer with three funnels. However, we achieved no success as we did not get close enough to the intended prey. We were compensated for this lack of success by the good news received by radio that the steamer "America", 22,000 tons, had run onto one of the mines we laid in the port of Halifax.[109]

The fighting mood was lifted again by this report, and we hoped confidently, despite the great difficulties, to bring down many more opponents. But there - what happened suddenly? Excited whispers from one to the other, mysterious whispers, shocked expressions, constant questions - in short, the entire crew was seized with great excitement when the rumor spread on October 21 that a radio message had been intercepted, in which we received orders to return home immediately and not attack any more vessels. The radioman was directed by the commander not to disseminate any of these orders. Nevertheless, it slowly tricked through on board that things were no longer right in Germany. Excited imaginations promoted all possible and impossible assumptions and conjectures. But nobody exactly knew. The poor radioman had no peace. He was constantly bombarded to tell us something. His cryptic comments were more confusing. What was going on in Germany? There was no other topic of conversation.

The only tangible fact was we were on a direct course home, and a few days later the well-known "shipyard layover time on the high seas" began again in order to bring the engines up to full performance for the dangerous return voyage. Alternately, the two diesel engines were dismantled and carefully checked. After more than a week, the engines were overhauled and we continued at full speed.

On November 7th we were again in front of the most dangerous part of our journey, the English minefield. Because of the lively sea we could not yet go through it. From 9 a.m. we lay quietly in front of

also witnessed all armed guard books and papers, and radio confidential books and papers dropped overboard."

the barrier and waited for better weather. But the next day, Friday, November 8th, the sea was even choppier. That's why we didn't dare to go through now. We turned back around again when a precise bearing of our position indicated we were already a few nautical miles in the field. But it was decided to try to pass the next morning under all circumstances.

We were all in a state of feverish excitement. From further leaked news we now knew at least the last war voyage was behind us. Now we were all tormented by anxiety: will we force the passage, or will we finally end up in the tragic fate of being ingloriously sunk? Despite all the uncertainty about the situation in Germany, there was joy the war was over. We just needed to get home again and start a new, quiet life again. But, but If only we got through this damned barrier.

On Saturday, 6.30 a.m. we set off again with both engines at emergency power and electrical power added as well, despite the still strong sea. And we were lucky. After a short time a submarine came into view - the German U 60. A few remarks were exchanged and then U 60 drove ahead. That was very reassuring to us. U 60 apparently knew the driving conditions better, and so we were able to safely pass the danger zone - always keeping close behind him.

On Monday, November 11th, 1918, our commanding officer decided it was time to inform us of the situation in Germany and of the fact that a republic had been proclaimed. We now knew the fears we had harbored for months had become true, and we now had to see how we could still successfully travel the rest of the way; because U 60 was completely out of sight again. The Sound was crossed with the greatest possible speed and also with extreme caution. There - we were still in the Sound, around 4 a.m., when suddenly a violent tremor went through the boat. The crew - already nervous in and of themselves - was seized with a sudden shock. But the calm commands of our leader quickly convinced us of the groundlessness of any fears. We just hit a sandbar and got stuck. However, this evil

---

[109] There is no record of this sinking. It may be planted disinformation.

had to be remedied. All the tanks were blown out and the boat lifted off so we could continue our journey.

At about 6 o'clock in the morning, on November 12th, we were in the area where the guard ship "Hannover" had been lying. But it was nowhere to be seen. From this circumstance we believed we could assume the mine fields laid out here had been pulled in and we could now drive through freely. But we were seriously mistaken in this assumption. We drove with the electric motors. I was on duty myself and in the control room. Suddenly we heard a strange rattle. Our boat had run into a mine cable. At the same moment "emergency astern" was ordered from the tower. In no time we had switched both engines, and with emergency power the boat backed up. At the last moment we were spared a terrible fate. A Swedish torpedo boat had warned our commander, who was on the bridge, at the last moment by telling him that we were running into our own mine field. Two or three seconds later, and we would have been blown up, we would have found our watery grave at the very last minute in our home waters. It was dangerous the mine barriers were simply left without a guard, although it was known a number of submarines were out and would be coming back. But the days of revolution had undermined any order.

Our next stop was Sassnitz on the island of Rügen. Various submarines were already lying here. As a result of rumors regarding very good conditions in Kiel, the commanders and officers could not make up their minds to go there. A soldiers' council was formed on every boat, as this was the norm everywhere. After waiting for two days in Sassnitz, the commanders and officers, together with the soldiers' councils, decided to go to Kiel. As we entered the harbor here, a steam launch came towards us, manned by armed sailors. One of them told our captain to immediately take down the battle flag. Captain Studt refused. He was then threatened with immediate execution if the flag was not removed as quickly as possible. Seeing there was nothing else to be done here, he had the flag drawn under which we had crossed the seas for years. That was our reception in Kiel. We all imagined a different homecoming.

Our captain then had the entire crew step up on deck again. Deeply moved, he said goodbye to us and thanked us for the faithful service which the crew had rendered in every situation, even in the most difficult. The fatherland, too, he remarked, owed us thanks for our service on the distant seas. But now, in these difficult days, we shouldn't expect to get the recognition we deserve. Only later would it be seen what deeds we had accomplished and what a tremendous achievement it was to fight for months in a small boat against an overpowering enemy. With tears in his eyes he now held out his hand to everyone. The firm handshake showed the warmth of his wish that we should all fare well.

After a few days our proud and successful merchant and war submarine "Deutschland" was handed over to the British.[110]

[110] U 155 was exhibited at London and Liverpool as a war trophy. The vessel was demilitarized and embarked on a commercial tour in 1919-1920; then laid up for stripping of valuable metals in 1921. In September 1921 an explosion of overlooked hydrogen gas tanks killed five apprentices. The hulk was sold in 1922 for breaking up.

Made in the USA
Monee, IL
04 November 2026

67384914R00075